T0257647

Geomorphology and Environment:
Analysis and Applications

Geomorphology and Environment:
Analysis and Applications

Edited by **Ken Shaw**

New York

Published by Callisto Reference,
106 Park Avenue, Suite 200,
New York, NY 10016, USA
www.callistoreference.com

Geomorphology and Environment: Analysis and Applications
Edited by Ken Shaw

International Standard Book Number: 978-1-63239-357-9 (Hardback)

Printed in the United States of America.

Contents

Preface

The purpose of the book is to provide a glimpse into the dynamics and to present opinions and studies of some of the scientists engaged in the development of new ideas in the field from very different standpoints. This book will prove useful to students and researchers owing to its high content quality.

Extensive information regarding the field of geomorphology and its applications as well as analysis has been compiled in this book. It provides numerous up-to-date studies on geomorphology, covering distinct disciplines and methodologies, aimed at tools, techniques and general issues of environmental and applied geomorphology. The aim of this book is to integrate various methodological areas (meteorological and climate analysis, geomorphological mapping, geographic information systems GIS, remote sensing, land management techniques, vegetation and bio-geomorphological investigations) spanning several countries across continents (Asia, Europe, Africa, America) across distinct study areas.

At the end, I would like to appreciate all the efforts made by the authors in completing their chapters professionally. I express my deepest gratitude to all of them for contributing to this book by sharing their valuable works. A special thanks to my family and friends for their constant support in this journey.

Editor

Biogeomorphologic Approaches to a Study of Hillslope Processes Using Non-Destructive Methods

Pavel Raška

Jan Evangelista Purkyně University in Ústí nad Labem,
Czech Republic

1. Introduction

The aim of this chapter is to present new non-destructive methods and techniques used in the biogeomorphologic study of hillslope processes, particularly sheet erosion and shallow landslides. These processes belong to a broad spectre of natural hazards that have significant impacts on landscape and society and their research represents the fundamental issue for applied geomorphology (Panizza, 1996; Alcántara-Ayala, Goudie eds., 2010). Non-destructive methods are not yet well established in biogeomorphologic research despite their relevance in areas protected under conservation law, in fragile habitats and considering their simple field application. To introduce some of these methods in case studies within the context of biogeomorphology, we first give an introduction to the main concepts regarding landform-biota interactions followed by a focus on hillslope processes. In sections 3 and 4, we present two case studies of the application of non-destructive methods to quantify the bioprotective role of fallen trees (trunk dams, log dams) and to analyse short-term surface stability. In the final section, we suggest possible directions for the future development of non-destructive methods in the biogeomorphology of hillslope processes.

The evolution of Earth's surface in contrast with other planets in the Solar System is characterised by the fundamental role played by organisms, which act directly by creating, modifying and destroying landforms and indirectly by changing other factors that influence surface processes, such as climate and the distribution of energy. Looking at the history of research in this field of expertise, it seems that the significance ascribed to organisms (and particularly to vegetation) within a short history of biogeomorphology grew as rapidly as other fundamental concepts within a hundred year history of geomorphology. Most recently, this trend has led to the assumption that vegetation can indeed be a leading factor in global geomorphic change, and understanding its evolution is crucial to establishing an evolutionary view in geomorphology (Corenblit & Steiger, 2009). From a case study in eastern Kentucky, for instance, Phillips (2009) concluded that if only 0.1 % of net primary production of biomass is assumed to be geologically active, it still exceeds the energy of uplift and denudation. In spite of these results, one can hardly imagine vegetation being responsible for creating the global geomorphic patterns comprising everything from mountain ranges to valleys (cf. Scheidegger, 2007) even though the absence or presence and character of vegetation can modify the rate of landform evolution by protecting the Earth's surface (factor being limited by extent of rhizosphere) or by changes in energy diversification (assuming land cover pattern scale, which is sufficient to influence continental to global climate). Moreover, the role of vegetation in different time horizons (i.e., scale dependency) is unclear. The classic work of Schumm & Lichty (1965) shows how vegetation changes from a dependent to independent factor through time. More recently, Phillips (1995), at a more local scale, has shown that vegetation may in fact be a dependent variable rather than a controlling factor, which has also been discussed by Raska & Orsulak (2009).

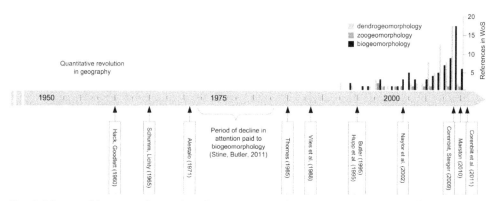

Fig. 1. Major publications devoted to the interaction of organisms and geomorphology shown in historical perspective.

The abovementioned uncertainties in geomorphic significance of organisms are partly caused by the development of the scientific background of biogeomorphology, the conceptualisation of which as a young subdiscipline of geomorphology is still developing. An overview of the main achievements in biogeomorphology as represented by fundamental publications is shown in Figure 1. The first works to discuss the landform-organism relation date back to the 19th century; however, it was not until 1960s that biogeomorphology was established as a research approach by Hack & Goodlett (1960). The very first attempt to give an overview of the field was in the 1980s, when Viles (in Viles et al. 1988) defined biogeomorphology as a "concept of an approach to geomorphology, which explicitly considers role of organisms". Since then, biogeomorphology has been increasingly pursued in studies across varying environments. According to the Web of Science database (Thomson-Reuters), the annual number of works with biogeomorphology as a focal topic increased from less than five to more than 15 during the last 20 years (Fig. 1). This trend is also apparent in dendrogeomorphologic studies, while the number of zoogeomorphologic papers remains consistently low. This pattern indicates the prevailing appreciation of vegetation, the role of which can be studied at different scales and can be generalised. In contrast to the increasing number of biogeomorphologic papers, it is somewhat surprising that the amount of biogeomorphologic content has decreased in textbooks, being lowest in 1970s (see Stine & Butler, 2011).

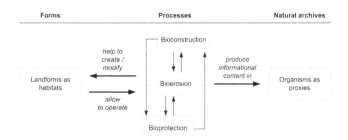

Fig. 2. The thematic framework of biogeomorphology (Naylor et al., 2002; modified by author)

The major aims of biogeomorphology were set by Viles et al. (1988) as (i) the influence of landforms on the distribution of organisms and (ii) the influence of organisms on Earth surface processes, but in fact, the two-way linkages were not fully pursued in the book (cf. Wainwright & Parsons, 2010). The "revision" of developments in biogeomorphology presented by Naylor et al. (2002) emphasises three main processes interlinking landforms and organisms: (i) bioconstruction, (ii) bioprotection and (iii) bioerosion. The authors call for studies of the complexity of landform-biota interactions. However, organisms were still considered a static factor, and geomorphologists were often unable to avoid the unidirectional appreciation of the landform-organism (vegetation) relation, which has changed only in past few years (e.g., Marston, 2010; Reinhardt et al., 2010; Corenblit et al., 2011).

The problem of a biogeomorphologic focus on two-way linkages is twofold. First, the development of biogeomorphology was accelerated as geomorphology moved from global and regional research scales aiming at historical interpretations of landscape (Church, 2010) to local scales and individual sites as a result of the quantitative revolution and the establishment of the process approach paradigm. The physical background of the process approach emphasised analyses of Earth surface processes and frequently neglected the feedbacks from organisms. Furthermore, a focus on a local scale did not allow the effective modelling of vegetation changes as a response to geomorphic processes because these changes are also influenced by decision-making effects, which are variable and can be better assessed at a regional scale (see Wainwright & Millington, 2010). One of the possible integrating views to resolve these scale-related problems could be offered by Quaternary landscape ecology, which emphasises evolutionary concepts together with changing patterns of biota and developments in human societies that influence the primary modes of decision making (e.g., Delcourt & Delcourt, 1988).

The second problem is methodological and emerges from the different backgrounds of the two disciplines integrated within biogeomorphology: geomorphology and ecology. While geomorphologic conclusions are frequently drawn from theoretical considerations combined with detailed field surveys and measurements, ecological conclusions usually result from statistical analyses of extensive datasets (Haussmann, 2011). The prevailing geomorphologic approach, then, only anticipates the real situation, where biota is frequently understood as a static factor or as a source of proxy data for the study of Earth surface processes. Fig. 2 shows the thematic framework of biogeomorphology emerging from

Naylor et al. (2002) and modified to emphasise the mutual linkages between biota and landforms.

2. Biogeomorphology of hillslope processes: Main themes and recent advances

The abovementioned constraints of a study of landform-biota interactions are quite apparent in the research of hillslope processes, as these represent one of the primary focuses of geomorphologists, and the evolution of biogeomorphology was tightly connected with advances in hillslope studies. Hillslopes represent the most common landforms across environments varying from periglacial to tropic regions (e.g., Anderson & Brooks, 1996 eds.), and they were a key landform in designing fundamental models of landscape evolution, ranging from the classic models of W.M. Davis, W. Penck and L.C. King (for overview see Summerfield, 1991) to modern ones, such as backwearing model by R.V. Ruhe, R.B. Daniels and J.G. Cady, nine-unit model by A.J. Conachre and J.B. Dalrymple, COSLOP by F. Ahnert, and many others. Much of the Earth's surface represented by hillslopes is covered by vegetation, whether a sparse cover of lichens, mosses, grass or less or a more continuous cover of shrubs and trees. Furthermore, the hillslopes are habitats for various animals, some of which prefer gentle slopes covered with a deep soil layer and others that developed a preference for rock-mantled slopes (see Butler, 1995, for overview). At the same time, hillslopes belong among the most dynamic landforms of the Earth's surface, enabling the occurrence and acceleration of different natural hazards, such as sheet and gully erosion, landslides, rockfalls, rock avalanches, debris flows, snow avalanches, and others, which have impacts on the manmade objects and human activities in a landscape (Alcántara-Ayala & Goudie, 2010 eds.). The enormous share of hillslopes on the Earth's surface together with their dynamics implies the necessity of intense applied research as well as opportunities for the development of new, effective research techniques. Finally, considering the abovementioned distribution of organisms on hillslopes, these techniques frequently draw upon analyses of organisms, whose distribution, activity or physiognomic modifications serve as proxy indicators of hillslope processes, and the primary attention is devoted to vegetation in this respect.

The first work on dendrogeomorphologic responses to geomorphic processes in terms of their chronology was by Alestalo (1971). Since then, many studies have focused mainly on mass movements and erosion. Concerning research on erosion, Thorns (1985) summarised not only the state of the art but also established a framework to understand feedbacks between vegetation and erosion, thus extending the traditional unidirectional approach in biogeomorphology; this was also enabled by his attention to non-linear dynamic systems in biogeomorphology. Most recently, Marston (2010) presents a comprehensive overview of research on hillslope-vegetation linkages, including research history, main functions of vegetation, feedbacks between vegetation and landforms within the disturbance regimes, and suggestions for future research directions.

An overview of the basic approaches and techniques used in the biogeomorphologic study of hillslope processes is presented in Table 1 along with references to some of the major papers published within this scope. Regarding the methods used, these approaches could be divided into three groups. The first one has in common the use of modelling techniques based on physical laws, which are quite often calibrated by the results of limited field surveys or laboratory measurements. These approaches are applied mostly in the modelling of changes in vegetation patterns and the related influence on sediment supply, and they often exploit specially developed GIS-based modelling software, such as MIKE 11 or HEC-RAS, although the incorporation of vegetation parameters into these models is limited.

Aim (process or effect)	Studied object (approach)	Reference
sheet erosion - detection, dating	roots (anatomy analyses)	Gärtner et al. (2001), Bodoque et al. (2005), Gärtner (2007), Rubiales et al. (2008)
sheet erosion - protective function of vegetation	vegetation (modelling)	Dorren et al. (2004), Vanacker et al. (2007)
gully erosion - detection, dating	roots (anatomy analyses)	Vandekerckhove et al. (2001)
shallow landslides - effect of roots on slope stability	roots (modelling)	Wu et al. (1979), Abe & Ziemer (1991), Preston & Crozier (1999), Schwarz et al. (2010)
shallow landslides - protective function of trees	forest cover (modelling)	Sidle et al. (1985), Bathurst et al. (2010)
landslides - detection, dating	trunk (tree ring analyses)	Fantucci (1999), Gers et al. (2001)
uprooting - effect on soils	roots, trunk (field survey)	Schaetzel et al. (1989), Phillips & Marion (2006)
rockfall - detection, dating	trunk (scars, tree ring analyses)	Perret et al. (2006), Stoffel & Perret (2006)
debris flows - detection, dating	trunk (scars, tree ring analyses)	Bollschweiler et al. (2008), Stoffel (2010), Šilhán & Pánek (2010)
protective effect of fallen trees - quantification	trunk (field survey, modelling)	Raska & Orsulak (2009)
effect of fallen trees - shaping debris flows	trunk (field survey)	Lancaster & Hayes (2003), Matyja (2007)
soil and regolith disturbances by animals - pattern, rate	different species (field survey and experiments)	Trimble & Mendel (1995), Gover & Poesen (1998), Hall & Lamont (2003)

Table 1. Overview of selected biogeomorphologic approaches to a study of hillslope processes (references are selected to show the different approaches and methodological advances in the field of expertise)

The second group of approaches can be called non-destructive. These approaches focus on measurements and analyses of visible features that represent past or current interactions between vegetation and hillslope processes. To analyse rockfall activity and patterns in protected forests, Stoffel (2005) performed analyses of the distribution and visibility of scars on trees. Attention was also given to log jams shaping debris flows trajectories (Lancaster et al., 2003; Matyja, 2007). However, most of these techniques are usually combined with sampling techniques in the field.

The third and most exploited group of approaches, beginning with the classic work of Alestalo (1971), draws upon extensive field research and sampling using different strategies. The samples are usually taken as cores, wedges or cross-sections from stems, stumps and roots (for nomenclature see Gschwantner et al., 2009). Sampling and tree ring analyses are frequently supported by measurements of stem or root deformation, and studies based on this approach have already offered good results regarding the spatiotemporal patterns of rockfall (e.g., Stoffel & Perret, 2006), debris flow (Bollschweiler et al., 2008) and landslide (e.g., Fantucci, 1999) activity. The application of the approach encounters several problems, however. These problems consist of the number and suitability of tree species for cross dating (Grissino-Mayer, 1993), the availability of reference datasets and other issues (see Stoffel & Perret, 2006). The number of samples differs according to the extent of the area under research, the time span and type of process being studied (Fig. 3), which result in the varying efficiency of the sampling strategy as expressed by the number of samples per one detected event.

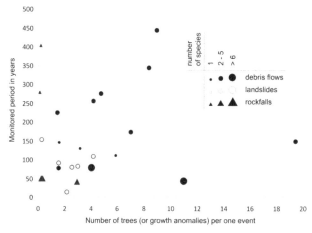

Fig. 3. Number of samples used for the detection and dating of debris flows, landslides and rockfall events in relation to the length of the monitored period in years and to the number of species analysed. Source: based on own review of 25 selected papers published in international journals between 2001 and 2011.

In some areas, however, it is not possible or suitable to take cross-sections and wedges from roots and stems or to measure root geometry deformations after uncovering the soil layer. In these cases, one has to employ non-destructive techniques, which enable the detection of processes and the estimation of their rates from the current positions and visible deformations of vegetation in the field. Moreover, several of the abovementioned destructive approaches focus the enormous potential of information recorded in vegetation species as proxy indicators, which oftentimes leads to maintaining the unidirectional approach to a study of landform-biota relations as discussed in the first section.

3. Structure, dynamics and protective effects of trunk dams

In this section, we will define and characterise trunk dams in terms of their structure, dynamics and protective effects. Then, we will present the recent advances in implementing the results of field research to regional models of the bioprotective effect of trunk dams at a catchment scale using the geographic information systems (GIS). The aim of the developed model is to transfer the local information about the protective effect of individual trunk dams to the regional scale and to estimate the total volume of material retained by fallen trees or trunks.

3.1 The concept

The major recent focus of biogeomorphologic (dendrogeomorphologic) studies of hillslope processes is on the protective effects of trees and shrubs. The research is especially carried out in protection forests. The delimitation of protection forests emerges from their general ability to control or modify the natural hazards connected with Earth's surface dynamics (Berger & Rey, 2004). Different national approaches and nomenclatures have been adopted for the zonation of protection forests and for their management, the difficulty of which emerges from the requirement to maintain both the ecosystem's integrity and the protective function of these forests (Dorren et al., 2004). The protection forests in mountainous areas are predominantly determined for rockfall impact reduction. Most protection forests in highlands and slightly undulating terrains without rock faces have a protective function against soil erosion. In the Czech Republic, where the present study was performed, these forests are called soil-protecting forests (Collective, 2007). In many studies and applied works on protection forests, the focus is on standing trees and shrubs, although Dorren et al. (2007), for instance, also explicitly mention the significance of lying trunks for increasing surface roughness and reducing the velocity of bouncing and rolling clasts of rock. The overall number of biogeomorphologic papers dealing with fallen trees, however, is small, and the issue has been much more in the focus of forest ecologists and biologists.

In forest ecology and management, fallen trees and their large parts are referred to as coarse woody debris (CWD), woody detritus, or downed wood, and a significant ecological role is ascribed to the process of tree death and decomposition (Masser et al., 1984; Harmon et al., 1986; Franklin et al., 1987). Lying trees and their parts increase the habitat diversity in forests, thus increasing biodiversity, as they offer good conditions for the development of mosses, lichens, plants and fungi communities as well as insect (*Insecta*) and other organisms. Furthermore, the importance of fallen trees lies in their influence on channel diversity in water streams, in the increase of the production function of forests and in the positive influence on the carbon cycle in forests. The ecological significance of fallen trees has led to changes in the traditional forest management approach, the aim of which formerly was the removal of all woody debris (cf. Harmon, 2002).

The trunk dam concept presented here was originally described in Raska & Orsulak (2009), resulting from the typology of the bioprotective effects of trees at rock-mantled slopes in the Ceske stredohori volcanic mountain range in the NW Czech Republic (Raska, 2007, 2010). In contrast to studies of allochtonous log jams, which often form as a result of mass movements on hillslopes, the present concept emphasises the role of individual trunks (or logs), which, depending on the local surface roughness and density of standing trees, are rather autochthonous. The trunk dam forms as the tree or its parts fall and are stabilised by obstacles on the surface. Depending on topography, density, the total length and other characteristics of trunk dams in the area, they can play a significant role in the reduction of the volume of material that is transported downslope. The collective (2007) shows that the amount of dead wood can be higher than 8 m^3.ha^{-1}. Vallauri et al. (2003) lists examples from European forests, which show that the volume of dead wood varies from 0.6 to almost 20 m^3.ha^{-1}. The total volume of fallen trees depends on the density and diversity of forest cover and on its ecological integrity.

Fig. 4. Trunk dam in advanced stage of plant succession (left), trunk dam breach (right).

3.2 Structure and dynamics

A trunk dam is a biogeomorphologic system compound of five components, three of which form the mass of the trunk dam and two of which (base surface, barriers of embedment of a trunk) create its boundaries. The three inherent components are the following (Figs. 4, 5C):

i.　Trunk: This component usually appears as a stem with large branches, but sometimes it can also be present in the form of a complete tree or as individual large branches. Depending on surface roughness and trunk geometry, the height of its position above the surface can vary.

ii.　Body: The main body of a trunk dam is formed by a mixture of sediments of varying fractions (from clay to angular scree and rounded boulders) and organic components including decomposed parts of plants, litter, etc. Its top can vary from flattened (cf. Matyja, 2007) to rough and inclined at locations, where the base surface has a higher inclination and where the trunk dam is subject to further impacts of rolling and bouncing clasts. The internal structure of the body depends on the evolutionary history of each trunk dam and the type of material, which is delivered from the upper parts of the hillslope. The bodies of the newly formed trunk dams include an unsorted mixture of material, but when the trunk dam is stabilised for longer time, the body can display stratification as a result of the sequence of events that delivered the material to the trunk dam and/or as a result of a sieve effect and sorting within the trunk dam body. An example of a trunk dam with such a partly stratified body as detected during the field survey (Raska & Orsulak, 2009) is shown in Fig. 5C.

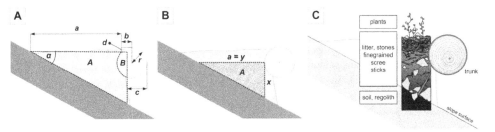

Fig. 5. Cross-section through the trunk dam. A - real (grey) and theoretical (black dashed) geometry of trunk dam with parameters measured in the field (lower case) and computed values (upper case); B - simplified trunk dam geometry used for spatial modelling (see section 3.3); C - structure of a trunk dam body.

iii. Cover: Trunk dams that have been stabilised long enough are covered by plants. The dynamic habitat of trunk dams is suitable for pioneer species. The results of a field survey in the study area show that the plants inhabiting trunk dams often belong to invasive species (e.g., *Impatiens glandulifera*). The speed of colonisation and the relative importance of species depend on the type of environment in which the trunk dam is located (McCullough, 1948). The fast colonisation of a trunk dam surface accelerates the stabilisation of its body through plant roots. While the body of a trunk dam is colonised mostly by herbs, the trunk itself in different stages of decay represents a habitat for mosses and especially fungi (e.g., Bader et al., 1995).

One of the aims of our previous research was to design a simple non-destructive method for estimating the volume of material accumulated by trunk dams. After initial attempts, an MS Excel Spreadsheet was designed. The EVAM (Estimation of Volume of Accumulated Material v1.0, available at <http://lsru.geography.ujep.cz>) approximates the cross-section of a trunk dam to a simple triangle-like shape (Fig. 5A), and it enables an accurate estimation of the volume of accumulated material (VAM).

The first results, which compared the accumulation and denudation rates in two sampling sites (Raska & Orsulak, 2009) showed that in terms of the current dynamics of a hillslope surface, the accumulation rate may be as important as the denudation rate. The overall results for 26 trunk dams analysed at three sites within the studied catchment revealed certain relations between the individual geometric parameters of a trunk dam (Fig. 6A). The highest correlation is between accumulation width "a" and VAM, but it will be shown later that there are certain difficulties in the generalisation of this relation. By contrast, a weak correlation was revealed between trunk radius "r" and VAM. However, the trunk radius and geometry of a trunk may partly influence other values, such as "a" and the height of a trunk above the surface.

The height of a trunk above the surface represents one of the most important factors influencing VAM values, as will be shown in relations between the width of accumulation, the slope inclination and VAM (Figs. 6B and 6C). Fig. 6B shows how VAM grows with increasing slope inclination. The model assumes that the height of a trunk above the surface is a dependent variable calculated from other input parameters. By contrast, the model in Fig. 6C assumes the trunk dam is in contact with the surface and, thus, is an independent variable. In this case, the higher is the slope inclination, the lower is the VAM.

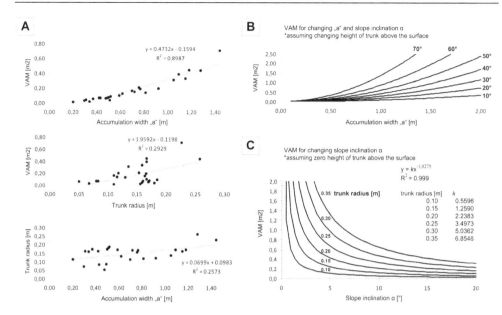

Fig. 6. A - relationship between measured parameters and the volume of accumulated material (VAM) in the dataset from a field survey; B, C - theoretical values of VAM and "a" for varying and for stable (zero) height of trunk above the surface (see text for further explanation).

As we discussed in section 3.1, the reason for the biogeomorphologic study of trunk dams is their ability to decelerate the velocity of rolling and bouncing clasts and their protective role against sheet erosion. The protective function against clasts, which move downslope as a result of rockfall events, is enabled by the trunk acting as a barrier and by the deceleration of clasts at the flattened top of a trunk dam (Raska, 2007). The bioprotective effect against sheet erosion is more complex, and its significance varies in the different parts of a trunk dam. In general, the bioprotective effects against sheet erosion could be direct and indirect. The direct effects consist of the accumulation of allochthonous material in the body of a trunk dam for a certain time. The indirect effects emerge from local feedbacks between trunk dam ecology and surface dynamics. As trunk dams are colonised by plants, they reduce soil erosion by the following:

- interception and deceleration of the velocity of raindrops falling on the ground;
- enhancing infiltration through root systems and increasing soil aggregate stability;
- transpiration of soil water, thus decreasing water content in soil;
- increasing surface roughness by roots, fallen leaves, etc.;
- increasing the volume of organic substances in the soil (Trimble, 1988; Gyssels et al., 2005).

However, the surface below the trunk dam often displays higher rates of denudation, as the trunk dam body retains all the sedimentary material. Also, the overland flow, which continues downslope, has higher erosive energy (see Fig. 4).

The fundamental point in the evaluation of the protective effect of trunk dams is the time scale in which they can effectively protect the surface against erosion. The stability of a trunk dam depends on several factors and can be expressed by the following equation:

$$S = f.(E_{s,e,d,p}, I_{p,w,m})$$ (1)

where:

- E ... external factors (s - slope inclination, e - embedment, d - disturbations, p - stabilisation by plants), and
- I ... internal factors (p - position of a trunk dam, w - wood characteristics, including strength, geometry and stage of decomposition, m - structure of material within the trunk dam body).

The role of these factors is summarised in Table 2. The persistence of a trunk dam is limited by the total decomposition of the trunk and emptying of the accumulated material or by the breach of a trunk dam. Nevertheless, in some cases, the body of a trunk dam can be stabilised by plants to such an extent that even the total decomposition of a trunk will not result in the destruction of a trunk dam (see the three-way development model in further text).

Factor	Stability	Instability
slope inclination	lower slope inclination	higher slope inclination
embedment	embedded by standing trees	embedded by surface roughness
disturbances	no impacts	rockfall, zooturbation
stabilisation by plants	with plant cover	without plant cover
position of a trunk dam	transverse	downslope
wood characteristics (strength, geometry and stage of decomposition)	large trunk radius, straight trunk, early stage of decomposition	small trunk radius, curved trunk, late stage of decomposition
structure of material within the trunk dam body	stratified, interconnected with base surface	not stratified, discontinuity between trunk dam body and base surface

Table 2. Factors influencing the stability of a trunk dam.

Related to the stability issues is the necessity to evaluate the persistence of the protective effect of trunk dams over time. It emerges from the very nature of the problem that the protection given by fallen trees, which are subject to gradual decomposition, will be effective only at small time scales, i.e., from the fall of the tree to its total decomposition or breach (Fig. 4). According to the results of the field survey, the age of individual trunk dams varied from a few days (newly fallen trees) to more than a year. The dating of a trunk dam can be performed by analysing the following indicators:

- the stage of trunk decomposition - large pieces of wood decay relatively slowly. Harmon et al. (1986) has shown that in temperate regions, half-time decay can vary from 25 yr (*Quercus*) to 150 yr (*Pseudotsuga*). Schowalter et al. (1992) conducted an experiment to assess the decomposition process of four species (*Pseudotsuga, Tsuga, Abies* and *Thuja*) and concluded that the decomposition is significantly influenced by the initial wood chemistry and by the colonisation pattern, especially the penetration of the bark barrier and colonisation by wood fungi. Harmon et al. (2000) have presented a new method for estimating biomass loss by wood decay. Their results have shown that in most cases, the biomass loss has negative exponential progress, while in one case (*Pinus sylvestris*), the regression trend was polynomial, displaying different phases of decomposition. The relative dating of trunk decomposition is usually made by distinguishing categories of (i) hard timber, (ii) edge soft - centre hard, (iii) edge hard - centre soft and (iv) totally rotten (e.g., Collective, 2007);

- the internal structure, stratification and compaction of the accumulated material (although this can also reflect initial processes that are responsible for the accumulation of material);
- the stage of plant succession on the top of the trunk dam body;
- the age, distribution and position of the wood fungi, which is to be discussed in section 4.

Based on the field survey, Raska & Orsulak (2009) proposed a hypothetical model of the three-way development of a trunk dam in the mid-segment of the hillslope. The three ways are the following: (i) stabilisation of the trunk dam by vegetation cover, (ii) dam breach and formation of a new trunk dam and (iii) denudation after the dam breach without the formation of a new trunk dam.

3.3 Modelling the protective effect at a catchment scale - A first glimpse

In the next step, after the field research and after designing the method for estimating the volume of accumulated material, our aim was to exploit the results in the regional model at a catchment scale, as simple regional modelling is the very effective way to implement the results into the decision making process. Until now, there have been several models developed for the quantification of erosional rates within small catchments, some of them mentioned earlier in the text. A comprehensive summary of these models is presented by Gyssels et al. (2005), with special attention paid to the role of vegetation in these models. All of these models confirm the increase of soil erosion with the decreasing percentage of vegetation cover.

Gyssels et al. (2005) also show that these models ascribe a protective role to biomass above the surface, but give little attention to the roots. Moreover, many of these models are designed for grasslands and agricultural plants, but the protective role of forests against soil erosion is much higher (Kirkby, 1980). As the monitoring and quantification of fallen trees in forest ecosystems is difficult, the traditional models incorporate variables related to living (standing) trees, i.e., to forest canopy structure. Therefore, we had to establish a new model to simulate the regional protective effects of trunk dams.

We used the GIS-based distributed (GRID-based) model that was applied to the experimental catchment, and the input values were set according to the results of the field survey. The first results of modelling were based on a simplified VAM equation implemented in the GIS environment (ESRI ArcView 3.2, ArcGIS 9.2).

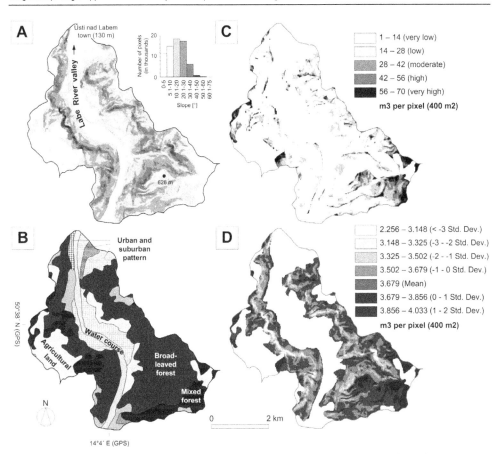

Fig. 7. Model of the protective effect of trunk dams within the experimental catchment. A - slope inclination derived from DEM (1:10000); B - land cover classes CORINE 2006; C - hypothetic calculated volume of accumulated material; D - calculated volume of accumulated material weighted by slope inclination (pixel size 20x20 m). Data sources: CORINE 2006 (European Environmental Agency), digital elevation data (CUZK).

The cross-section of a body of a trunk dam was assumed to be in contact with a trunk and having the shape of a rectangular triangle (Fig. 5B). Thus, the only necessary values for the calculation of VAM were the accumulation width "a", the length of trunk "L" and the slope inclination "α". Instead of trunk radius "r", we set the average "x" value (Fig. 5B) calculated from the field survey results. The variability of this value depends on the forest structure and age within the modelled catchment and on the surface roughness influencing height of a trunk above the surface. In the present study, the structure and age of forests is relatively uniform (Fig. 7B). There are no significant differences in trunk radius variations among the studied sites, and therefore, we set the uniform "x" value (0.35 m). Similarly, the length of trunks was set as an average from the field survey, which was 2500 m.ha⁻¹. Thus, the results of the simplified VAM model will correspond only to the expression of slope inclination. The slope inclination was derived from a digital elevation model (DEM; Fig. 7A). The pixel size of distributed model was set to 400 m^2, which reflects the precision of the input digital data (cf. Hengl, 2006). As the accumulation width is calculated from slope inclination and the "x" value, it can reach high values in locations with flat terrain; but, in fact, the accumulation width is always limited by the neighbouring trunk dams. The maximum accumulation width can be derived from a simple equation assuming the regular distribution of trunk dams:

$$\max a = PS^2/tL \qquad (2)$$

where PS is pixel size (i.e., 20 m for a pixel area of 400 m^2), and tL is the total length of trunks per pixel (i.e., 100 m). The result is a maximum accumulation width of 4 m, which was the value put as an upper limit into the model before acquiring the results of VAM.

The results of the modelling are shown in Fig. 7C, indicating that VAM varies between 1 - 70 m^3 per 400 m^2, i.e., 25 - 1750 m^3.ha^{-1}, but these values seem to be far from the real situation. The reason is that the "x" value was set constant, and therefore, the model assumes that the lower the slope inclination, the higher the accumulation width (see above) and VAM (see also Fig. 6C). In fact, the real processes operating on low gradient slopes will hardly enable accumulation of the material in the trunk dam with a similar efficiency as on highly inclined slopes. Evidently, there are other factors that influence the potential of trunk dams to be filled with accumulated material, such as the length of slope above the trunk dam, the surface material on the slope, and the disturbance regimes affecting the movement of material (e.g., overland flow, zoodisturbances, forest management measures). However, the slope inclination tends to be the most important of these factors according to our observations. The model (Fig. 7C) was therefore weighted again by the slope inclination to give higher importance to slopes with a high gradient and vice versa. The results of VAM in Fig. 7D vary between 2.2 - 4.0 m^3 per 400 m^2, which corresponds quite well to the empirical values gained during the field survey. Nevertheless, the development of the model is still in progress, and the results may differ when computed for more variable input values.

4. Interpreting active hillslope processes using polypore species

The indication and dating of hillslope processes connected with biotic communities is limited by the suitability of species (see in the previous sections) and by the time scale pursued by the research. Whereas the former applications of dendrochronologic dating focused on the identification of processes that occurred on a scale of years to hundreds of years (Stoffel et al., 2005; for overview see Fig. 3), and effort has been made to extend the time-span of reference datasets, the dating of current dynamics encounters several problems. Aside from minimal time, which is necessary for tree-ring growth, there are further constraints regarding the differences in the wood anatomy of different species and the limits of dating decomposed trunks. Dating at a time scale of months to a few years, therefore, calls for other proxy data.

An opportunity is given by a study of the succession of organisms at newly established habitats that have been formed as a result of Earth surface dynamics (e.g., Beschel, 1961; Pérez, 2010). Considering hillslope processes, these habitats may include log jams and trunk dams, which are colonised by various species (Masser et al., 1984). In this section, we focus on polypores (*Polyporales*).

Polyporous fungi belong to dead-wood-dependent organisms, and they contribute to the continuity of forest renewal by accelerating wood decay processes. The fungi-accelerated wood decay, in turn, improves the integrity of forest ecology and, therefore, the ecology of polypores must be reflected in forest management approaches (Lindblad, 1998). The diversity and abundance of polyporous fungi depends on the species diversity of trees, forest connectivity and the number of downed logs (Edman & Jonsson, 2001; Hottola, et al. 2009). Species richness also plays a role in the decay stage of wood and tree diameter, although the latter relation is a current topic of discussion (Junninen & Komonen, 2011). Although some polypores are able to establish ectomycorrhizal relationships, most polypore species are wood-dependent parasites, saprotrophs and necrotrophs. The reproduction principles and their relation to life histories and the diversity of polypore species are well discussed by Kauserud et al. (2008).

Specie	Habitat
Climacocystis borealis (Fr.) Kotl. & Pouzar	conifers
Trichaptum abietinum (Dicks.) Ryvarden	conifers
Fomitopsis pinicola (Sw.) P. Karst.	both conifers and broad-leaved trees
Chondrostereum purpureum (Pers.) Pouzar	both conifers and broad-leaved trees
Ganoderma applanatum (Pers.) Pat.	mostly broad-leaved trees
Daedalea quercina (L.) Pers.	broad-leaved trees (especially *Quercus*)
Daedaleopsis confragosa (Bolton) J. Schröt.	broad-leaved trees
Fomes fomentarius (L.) J. Kickx f.	broad-leaved trees
Phellinus igniarius (L.) Quél.	broad-leaved trees
Plicaturopsis crispa (Pers.) D. A. Reid	broad-leaved trees
Trametes hirsuta (Wulfen) Pilát	broad-leaved trees

Table 3. Selected Central European polypore species suitable for the indication of log jam dynamics and their usual habitats.

The fundamental importance for biogeomorphologic studies is represented by the basic ecological and morphological characteristics of polypores, especially the following;

- relation of polypores to concrete tree species (relation to conifers or broad-leaved trees - see Table 3 for overview of selected polypore species);
- colonisation rates on downed wood;
- maximal density of individuals;
- their position on downed wood (horizontal growth of individuals, groups);
- age of individuals - the age of a log jam may be deduced from the largest individual similar to the techniques used in the lichenometric dating of rock surfaces (Beschel, 1961), but at smaller time scales;
- growth rates (annual and perennial, visibility of increments);
- morpholgy of individuals and their deformations (tilting, rotation).

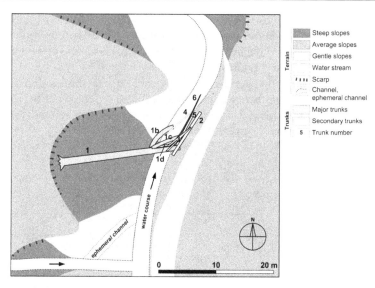

Fig. 8. Geomorphologic sketch of the study site showing the positions of trunks in the channel of a small water stream.

Polypores, which are suitable for the interpretation of the evolutionary history of log jams and fallen trees, predominantly grow horizontally with detectable increment rates. The indication of hillslope processes (erosion, shallow landslides) that affected the log jams and fallen trees is then allowed by the identification of the deformations of individuals. The most common deformations are tilting from the horizontal position and abnormal increments caused by the rotation of the host wood.

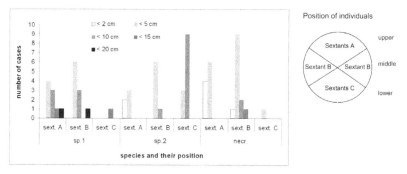

Fig. 9. Position of polypores on trunk number 1 and 1c (Fig. 8). Sp. 1 - *Fomitopsis sp.*, sp. 2 - *Daedalea sp.*, necr. - decomposed individuals.

We have carried out a detailed case study of the dynamics of a log jam in a small water stream and compared it with other examples from Central European forests. Fig. 8 presents a geomorphologic sketch of the study site with a log jam in the channel. The log jam was formed as a result of tree downing induced by lateral erosion in a deeply incised channel. After the major trunk was embedded in the channel, it stopped other allochthonous coarse woody debris. To understand the evolutionary history of the log jam, we analysed the distribution of polypore individuals on the major trunks located in the channel. The results indicate that different polypore species are located in variable positions within the trunk both in terms of the cross-section position through the trunk (Fig. 9) and of the position along the trunk. The typical morphology of the present species is depicted in Fig. 10. The deformations of individuals and the relation of different types of species and of living and decomposed polypores to positions within the trunk allowed us to specify the major events in the dynamics of the log jam.

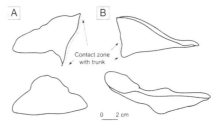

Fig. 10. Front and side views of the polypores on trunk number 1 and 1c. A - *Fomitopsis sp.*, B - *Daedalea sp.*

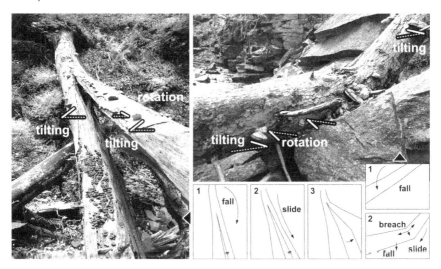

Fig. 11. Interpretation of the evolutionary history of two log jams based on polypore anatomy. The actual position after disturbance is shown by the white line and the hypothetic horizontal position by white dashed line. The black and white schemes in the right bottom depict evolutionary histories of the two log jams.

While the minimal height of living individuals above the water surface was approximately 0.5 m, the dead individuals were frequently located in the contact zone with the water surface. The presence of old decomposed polypores in an unsuitable position in contact with the water surface indicates the occurrence of an event that moved the trunk downward. The tilting and rotation of individuals caused by the movement of the host trunk is shown in Fig.

11, which depicts sample trunk number 1 (Fig. 11 left; see position in Fig. 8) and a comparative case of log jam from Western Carpathians (Fig. 11 right). The reconstruction of evolutionary history based on the identification of tilting and rotation determined the main processes that play a role in consecutive stages of log jam evolution and the interrelation with the surface morphology and dynamics of the locality. These processes include the initial fall of the tree, sliding, rotation and trunk breach.

5. Future directions of non-destructive biogeomorphology at hillslopes

Biogeomorphology is presently one of the most progressive branches of geomorphology, and thanks to its interdisciplinarity, it easily finds several issues in common with ecology, geobiology, and forest management, thus contributing significantly to the complex appreciation of Earth surface systems. The traditional unidirectional approach to a study of landform-biota interaction has gradually extended to cover the two-way linkages and feedbacks between Earth surface processes and landforms on the one hand and organisms on the other. Along with theoretical considerations, many studies show the feedback at concrete localities by employing different techniques. The biogeomorphologic research of hillslope processes is, however, traditionally engaged with destructive methods, such as dendrogeomorphologic sampling and exposure. These methods have been widely exploited in spatiotemporal analyses of mass transport in both high-altitude (e.g. rockfall, debris flows, etc.) and mid-altitude (e.g. landslides and sheet erosion) environments. The studies were carried out in order to understand regional geomorphologic effects of global environmental change (e.g. Evans, Clague, 1994) as well as to provide a scientific background for applied management measures (Panizza, 1996). On the other hand, the non-destructive techniques have been rather overlooked in this respect.

Non-destructive methods offer at least three important advantages, however. First, they can be applied in areas where destructive techniques cannot be performed. Second, they enable the performance of relatively fast and extensive field research in contrast to destructive sampling strategies, even though their results may not be as comprehensive, and there are still problems, which can be studied only by means of destructive sampling strategies (e.g., dendrochronology). Finally, these methods enable the study of some of those landform-biota interactions, which have been neglected. In some cases, these methods may also represent the only effective technique to study these interactions (e.g., short-term surface dynamics). In this respect, we see opportunities that should be more deeply explored by geomorphologists and which, in turn, certify the need for a bidirectional framework of biogeomorphology by employing ecological, ethological and biological knowledge. To name just one of the topics in which biogeomorphologists may more intensively draw upon non-destructive techniques, we would like to mention zoodisturbances at hillslopes.

As was mentioned in the first section, zoogeomorphology remains a "little sister" of dendrogeomorphology. Butler (1995) has shown the influences of the wide spectre of animals on Earth surface processes and landforms; however, only a few of these influences are studied systematically. The main focus has been burrowing animals (Hole, 1981; Volsamber & Veen, 1985) and animal trampling at rock-mantled slopes and in grasslands. The studies performed usually engaged experiments with manipulated animal density (e.g., Govers & Poesen, 1998) or detailed field mapping and soil sampling to reveal the physical and chemical properties of grazed land (Cluzeau et al., 1992; Trimble & Mendel, 1995). Despite producing important information about the rates of animal induced surface dynamics, these studies are scale dependent because their methodical background is time-exhaustive (detailed mapping and experiments). The opportunity to generalise the results of these studies lies in drawing from ethological information about the ecological behaviour of the studied species.

The example for consideration will be given from the study area, which was presented in the case studies in this chapter. The forest ecosystems of the studied catchment are diversified by patches of rock-mantled slopes and talus scree deposits, which have formed by the disintegration of rock-cliffs built by basaltic rocks since the Late Glacial Period. There are more than 100 such talus slope deposits exceeding an individual area of 400 m². During the study of these deposits, we focused on different types of processes, among which zoodisturbances tend to play an important role. Nevertheless, the monitoring of animal trampling on talus slope deposits exceeded the possibility of the research, so the other approach had to be adopted. The major trampling specie in the area is mouflon (*Ovis musimon*), which is an introduced specie in the Czech Republic (Heroldova & Homolka, 2001). As shown by Cransac & Hewison (1997), the seasonal activity and selection of habitats of mouflon hordes in their original environment is dependent on several variables, such as feeding activity and climate. During the year, the hordes are partly bound to rocky habitats. In our study area, it was confirmed by observation that talus slope deposits indeed represent an alternative habitat for mouflon hordes. The information about mouflon ethology together with the observation at talus slope deposits, where the presence of mouflon hordes was confirmed, enabled the specification of results from the field mapping of microtopographic features (especially clast flows) present on talus slope deposits (Raska, 2010). A similar approach could be applied to reveal the spatial behaviour and specific effects of other trampling species or burrowing animals.

To summarise, we are aware that there are many biogeomorphologic studies that are relatively close to non-destructive approaches, but were not mentioned in this chapter due to its limited extent. We refer readers to more complete reviews of biogeomorphology, dendrogeomorphology, and zoogeomorphology as well as the problem of soil erosion, which is too broad to be discussed in detail herein. The main aim of the chapter was to emphasise the importance of non-destructive biogeomorphologic approaches for a better understanding of the interwoven relationships between the Earth surface and organisms and to document this with two methods, which are currently being developed and applied at the catchment scale within the presented case studies.

6. Acknowledgements

Certain parts of the research were performed thanks to financial support from the research project IGA UJEP Disturbance regimes in the Quaternary morphogenesis of the Elbe river valley in the central part of the Ceske stredohori Mts. The author would like to thank Language Editing Services for the English style revision.

7. References

Abe, K. & Ziemer, R.R. (1991). Effect of Tree Roots on Shallow-Seated Landslides. *USDA Forest Service Gen. Technical Report PSW-GTR*, Vol.130, pp. 11-20

Alcántara-Ayala, I. & Goudie, G. eds. (2010). *Geomorphological hazards and disaster prevention.* Cambridge University Press, Cambridge

Alestalo, J. (1971). *Dendrochronological interpretation of geomorphic processes.* Vol.105. Soc. Geographica Fennia, Helsinki

Anderson, M.G.; Brooks, S.M. eds. (1996). *Advances in hillslope processes,* 1, 2. John Wiley & Sons., Inc., New York

Bader, P.; Jansson, S. & Jonsson, B.G. (1995). Wood-inhabiting fungi and substratum decline in selectively logged boreal spruce forest. *Biological Conservation,* Vol.72, pp. 355-362

Bathurst, J.C.; Bovoloa, C.I. & Cisneros, F. (2010). Modelling the effect of forest cover on shallow landslides at the river basin scale. *Ecological Engineering,* Vol.36, pp. 317-327

Berger, F. & Rey, F. (2004). Mountain protection forests against natural hazards and risks: new French developments by integrating forests in risk zoning. *Natural Hazards,* Vol.33, pp. 395-404

Beschel, R.E. (1961). *Dating rock surfaces by lichen growth and its application to glaciology and physiography.* In: *Geology of the Arctic, 2,* G.O. Raash, (Ed.), 1044-1062, University of Toronto Press, Toronto

Bodoque, J.M.; D´ıez-Herrero, A.; Martín-Duque, J.F., Rubiales, J.M.; Godfrey, A.; Pedraza, J.; Carrasco, R.M. & Sanz, M. A. (2005). Sheet erosion rates determined by using dendrogeomorphological analysis of exposed tree roots: two examples from Central Spain. *Catena,* Vol.64, pp. 81-102

Bollschweiler, M.; Stoffel, M. & Schneuwly, D.M. (2008). Dynamics in debris-flow activity on a forested cone – A case study using different dendroecological approaches. *Catena,* Vol.72, pp. 67-78

Brang, P.; Schönenberger, W.; Ott, E. & Gardner, B. (2001). Forests as protection from natural hazards, In: *The forest handbook,* J. Evans, (Ed.), 53-81, Wiley & Sons, Inc., New York

Butler, D.R. (1995). *Zoogeomorphology: animals as geomorphic agents.* Cambridge University Press, New York

Church, M. (2010). The trajectory of geomorphology. *Progress in Physical Geography,* Vol.34, pp. 265-286

Cluzeau, D.; Binet, F.; Vertes, F.; Simon, J.C.; Riviere, J.M. & Trehen, P. (1992). Effects of intensive cattle trampling on soilplant-earthworms system in two grassland types. *Soil Biology and Biochemistry,* Vol.24, pp. 1661-1665.

Collective (2007). *National forest inventory in the Czech republic 2001-2004: Introduction, Methods, Results.* UHUL, Brandýs nad Labem (Czech Republic)

Corenblit, D. & Steiger, J. (2009). Vegetation as a major conductor of geomorphic changes on the Earth surface: toward evolutionary geomorphology. *Earth Surface Processes and Landforms,* Vol.34, pp. 891-896

Corenblit, D.; Baas, A.C.W.; Bornette, G.; Darrozes, J.; Delmotte, S.; Francis, R.A.; Gurnell, A.M.; Julien, F.; Naiman, R.J. & Steiger, J. (2011). Feedbacks between geomorphology and biota controlling Earth surface processes and landforms: A review of foundation concepts and current understandings. *Earth-Science Reviews,* Vol.106, pp. 307-331

Cransac, N. & Hewison, A.J.M. (1997) Seasonal use and selection of habitat by mouflon (Ovis gmelini): Comparison of the sexes. *Behavioural Processes,* Vol.41, pp. 57-67

Delcourt, H.R. & Delcourt, P.A. (1988). Quaternary landscape ecology: Relevant scales in space and time. *Landscape Ecology*, Vol. 2, pp. 23-44

Dorren, L.K.A.; Berger, F.; Imeson, A.C.; Maier, B. & Rey, F. (2004). Integrity, Stability and Management of Protection Forest in the European Alps. *Forest Ecology and Management*, Vol.195, pp. 165-176

Dorren, L.K.A.; Berger, F.; Jonsson, M.; Krautblatter, M.; Mölk, M.; Stoffel, M. & Wehrli, A. (2007). State of the art in rockfall – forest interactions. *Schweiz Z. Forstwes.*, Vol.158, pp. 128-141

Edman, M. & Jonsson, B.G. (2001).Spatial pattern of downed logs and wood-decaying fungi in an old-growth *Picea abies* forest. *Journal of Vegetation Science*, Vol.12, pp. 609-620

Evans, S.G. & Clague, J.J. (1994) Recent climatic change and catastrophic geomorphic processes in mountain environments. *Geomorphology*, Vol.10, pp. 107-128

Fantucci, R. (1999). Dendrogeomorphology in landslide analysis. In: *Floods and landslides: integrated risk assessment*, R. Casale & C. Margottini, (Eds.), 69-81, Springer-Verlag, Berlin

Franklin, J.F.; Shugart, H.H. & Harmon, M.E. (1987). Tree death as an ecological process. *Bioscience*, Vol.37, pp. 550-556

Gärtner, H. (2007). Tree roots—Methodological review and new development in dating and quantifying erosive processes. *Geomorphology*, Vol.86, pp. 243-251

Gärtner, H.; Schweingruber, F.H. & Dikau, R. (2001). Determination of erosion rates by analyzing structural changes in the growth pattern of exposed roots. *Dendrochronologia*, Vol.19, pp. 81-91

Gers, E.; Florin, N.; Gärtner, H.; Glade, T.; Dikau, R. & Schweingruber, F.H. (2001). Application of shrubs for dendrogeomorphological analysis to reconstruct spatial and temporal landslide movement patterns. A preliminary study. *Zeitschrift für Geomorphologie*, Vol. 125 (Supplement), pp. 163-175

Govers, G. & Poesen, J. (1998). Field experiments on the transport of rock fragments by animal trampling on scree slopes. *Geomorphology*, Vol.23, pp. 193-203

Grissino-Mayer, H.D. (1993). An updated list of species used in tree-ring research. *Tree-Ring Bulletin*, Vol.53, pp. 17–43

Gschwantner, T.; Schadauer, K.; Vidal, C.; Lanz, A.; Tomppo, E.; di Cosmo, L.; Robert, N.; Englert Duursma, D. & Lawrence, M. (2009). Common Tree Definitions for National Forest Inventories in Europe. *Silva Fennica*, Vol.42, pp. 303-321

Gyssels, G.; Poesen, J.; Bochet, E. & Li, Y. (2005). Impact of plant roots on the resistance of soil to erosion by water: a review. *Progress in Physical Geography*, Vol.29, pp. 189-217

Hack, J.T. & Goodlett, J.C. (1960). *Geomorphology and forest ecology of a mountain region in the Central Appalachians.* U.S. Geological Survey Professional Paper 347.

Hall, K. & Lamont, N. (2003). Zoogeomorphology in the Alpine: some observations on abiotic–biotic interactions. *Geomorphology*, Vol.55, pp. 219–234

Harmon, M.E. (2002). Moving towards a new paradigm for woody Detritus Management. *USDA Forest Service Gen. Tech. Rep.*, PSW-GTR-181., pp. 929-944

Harmon, M.E.; Franklin, J.F.; Swanson, F.J.; Sollins, P.; Gregory, S.V.; Lattin, J.D.; Anderson, N.H.; Cline, S.P.; Aumen, N.G.; Sedell, J.R.; Lienkaemper, G.W.; Cromack, K. & Cummins, K.W. (1986). Ecology of coarse woody debris in temperate ecosystems. *Advances in Ecological Research*, Vol.15, pp. 133-302

Harmon, M.E.; Krankina, O.N. & Sexton, J. (2000). Decomposition vectors: a new approach to estimating woody detritus decomposition dynamics. *Canadian Journal of Forest Research*, Vol.30, pp. 76-84

Hausmann, N.S. (2011). Biogeomorphology: understanding different research approaches. *Earth Surface Processes and Landforms*, Vol.36, pp. 136-138

Hengl, T. (2006). Finding the right pixel size. *Computers & Geosciences*, Vol.32, pp. 1283-1298

Heroldova, M. & Homolka, M. (2001). The introduction of mouflon into forest habitats: a desirable increasing of biodiversity? In: *Proceedings of the third international symposium on mouflon*. Sopron, Hungary, pp. 37–43

Hole, F.D. (1981). Effects of animals on soil. *Geoderma*, Vol.25, pp. 75-112

Hottola, J.; Ovaskainen, O. & Hanski, I. (2009). A unified measure of the number, volume and diversity of dead trees and the response of fungal communities. *Journal of Ecology*, Vol.97, pp. 1320-1328

Hupp, C.R.; Osterkamp, W.R. & Howard, A.D. (1995). Biogeomorphology - terrestrial and freshwater systems. Proceedings of the 26th Binghamton Symposium in Geomorphology. *Geomorphology*, Vol.13, 347 p.

Junninen, K. & Komonen, A. (2011). Conservation ecology of boreal polypores: a review. *Biological Conservation*, Vol.144, pp. 11-20

Kauserud, H.; Colman, J.E. & Ravarden, L. (2008). Relationship between basiodiospore size, shape and life history characteristics: a comparison of poylpores. *Fungal Ecology*, Vol.1, pp. 19-23

Kirkby, M.J. (1980). *The problem*. In: *Soil erosion*, M.J. Kirkby & R.P.C. Morgan, (Eds.), 1-16, John Wiley & Sons., Inc., New York

Lancaster, S.T.; Hayes, S.K. & Grant, G.E. (2003). Effects of wood on debris flow runout in small mountain watersheds. *Water Resources Research*, Vol.39, pp. 1-21

Lindblad, I. (1998).Wood-inhabiting fungi on fallen logs of Norway spruce: relations to forest management and substrate quality. *Nordic Journal of Botany*, Vol.18, pp. 243-256

Marston, R.A. (2010). Geomorphology and vegetation on hillslopes: interactions, dependencies, and feedback loops. *Geomorphology*, Vol.116, pp. 206–217

Masser, Ch.; Trappe, J.M.; Cline, S.P.; Cromack, K.; Blaschke, H.; Sedell, J.R. & Swanson, F.J. (1984). *The seen and unseen world of a fallen tree*. General Technical Report PNW-164. U.S. Department of Agriculture, Portland

Matyja, M. (2007). The Significance of Trees and Coarse Woody Debris in Shaping the Debris Flow Accumulation Zone (North Slope of the Babia Gora Massif, Poland). *Geographia Polonica*, Vol.80, pp. 83-100

McCullough, H.A. (1948). Plant Succession on Fallen Logs in a Virgin Spruce-Fir Forest. *Ecology*, Vol.29, pp. 508-513

Naylor, L.A.; Viles, H.A. & Carter, N.E.A. (2002). Biogeomorphology revisited: looking towards the future. *Geomorphology*, Vol.47, pp. 3-14

Panizza, M. (1996). *Environmental geomorphology*. Elsevier, Amsterdam.

Pérez, F.L. (2010). Biogeomorphic relationships between slope processes and globular Grimmia mosses in Haleakala's Crater (Maui, Hawaii). *Geomorphology*, Vol.116, pp. 218-235

Perret, S., Stoffel, M. & Keinholz, H. (2006). Spatial and temporal rockfall activity in a forest stand in the Swiss Prealps - a dendrogeomorphological case study. *Geomorphology*, Vol.74, pp. 219-213

Phillips, J. D. (1995). Biogeomorphology and landscape evolution: the problem of scale. *Geomorphology*, Vol.13, pp. 337-347

Phillips, J.D. (2009).Biological energy in landscape evolution. *American Journal of Science*, Vol.309, pp. 271-289

Phillips, J.D. & Marion, D.A. (2006). Biomechanical effect of trees on soil and regolith: beyond treethrow. *Annals of the Association of American Geographers*, Vol.96, pp. 233-247

Preston, N.J. & Crozier, M.J. (1999). Resistance to shallow landslide failure through rootderived cohesion in east coast hill country soils, North Island, New Zealand. *Earth Surface Processes and Landforms*, Vol.24, pp. 665-675

Raska, P. (2007). Comments on the recent dynamics of scree slopes in the Czech Middle Mountains. *Geomorphologia Slovaca et Bohemica*, Vol.1, pp. 43-49

Raska, P. (2010). Types and character of geomorphic processes at Central-European low-altitude scree slope (NW Czechia) and their environmental interpretation. *Moravian Geographical Reports*, Vol. 18, pp. 30-38

Raska, P. & Orsulak, T. (2009). Biogeomorphic effects of trees on rock-mantled slopes: searching for dynamic equilibrium. *Geograficky casopis*, Vol.61, pp. 19-28

Reinhardt, L.; Jerolmack, D.; Cardinale, B.J.; Vanacker, V. & Wright, J. (2010). Dynamic interactions of life and its landscape: feedbacks at the interface of geomorphology and ecology. Earth Surface Processes and Landforms, Vol.35, pp. 78-101

Rubiales, J.M.; Bodoque, J.M.; Ballesteros, J.A. & Diez-Herrero, A. (2008). Response of Pinus sylvestris roots to sheet-erosion exposure: an anatomical approach. *Natural Hazards Earth System Sciences*, Vol.8, pp. 223–231

Schaetzel, R.J.; Johnson, D.L.; Burns, S.F. & Small, T.W. (1989). Tree uprooting: review of terminology, process, and environmental implications. *Canadian Journal of Forest Research*, Vol.19, pp. 1-11

Scheidegger, A.E. (2004). *Morphotectonics*. Springer, Berlin-Heidelberg

Schowalter, T.D.; Caldwell, B.A.; Carpenter, S.E.; Groffiths, R.P.; Harmon, M.E.; Ingham, E.R.; Kelsey, R.G. Lattin, J.D. & Moldenke, A.R. (1992). *Decomposition of Fallen Trees: Effects of Initial Conditions and Heterotroph Colonization Rates*. In: *Tropical Ecosystems: Ecology and Management*, K.P. Singh and J.S. Singh, (Eds.), 373-383, Wiley Eastern Limited, New Delhi

Schumm, S.A. & Lichty, R.W. (1965). Time, space and causality in geomorphology *American Journal of Science*, Vol.263, pp. 110-119

Schwarz, M.; Preti, F.; Giadrossich, F.; Lehmann, P. & Or, D. (2010). Quantifying the role of vegetation in slope stability: A case study in Tuscany (Italy). *Ecological Engineering*, Vol.36, pp. 285–291

Sidle, R.C.; Pearce, A.J. & O'Loughlin, C.L. (1985). *Hillslope stability and land use. Water Resources Monograph*, vol. 11. American Geophysical Union, Washington, DC

Silhan, K. & Panek, T. (2010). Fossil and recent debris flows in medium–high mountains(Moravskoslezské Beskydy Mts, Czech Republic). *Geomorphology*, Vol.124, pp. 238-249

Stefanini, M.C. (2004). Spatio-temporal analysis of a complex landslide in the Northern Apennines (Italy) by means of dendrochronology. *Geomorphology*, Vol.63, pp. 191-202

Stine, M.B. & Butler, D.R. (2011). A content analysis of biogeomorphology within geomorphology textbooks. *Geomorphology*, Vol.125, pp. 336-342

Stoffel, M. (2005).Assessing the vertical distribution and visibility of rockfall scars in trees. *Schweiz. Z. Forstwes.* Vol.156, pp. 195-199

Stoffel, M. (2010). Magnitude–frequency relationships of debris flows – A case study based on field surveys and tree-ring records. *Geomorphology*, Vol.116, pp. 67-76

Stoffel, M.; Schneuwly, D.; Bollschweiler, M.; Lievre, I.; Delaloye, R.; Myint, M. & Monbaron, M. (2005). Analyzing rockfall activity (1600-2002) in a protection forest—a case study using dendrogeomorphology. *Geomorphology*, Vol.68, pp. 224-241

Stoffel, M. & Perret, S. (2006). Reconstructing past rockfall activity with tree rings: Some methodological considerations. *Dendrochronologia*, Vol.24, pp. 1-15

Summerfield, M.A. (1991). *Global geomorphology. An introduction to the study of landforms.* John Wiley & Sons., Inc., New York

Trimble, S.W. (1988). *The impact of organisms on overall erosion rates within catchments in temperate regions.* In: *Biogeomorphology*, H. Viles, (Ed.), 83-142, Basil Blackwell, Oxford

Trimble, S.W. & Mendel, A.C. (1995). The cow as a geomorphic agent - A critical review. *Geomorphology*, Vol. 13, pp. 233-253

Vallauri, D.; André, J. & Blondel, J. (2003). Le bois mort, une lacune des forêts gérées. *Revue Forestière Française*, Vol.2, pp. 99-112

Vanacker, V.; von Blanckenburg, F.; Govers, G.; Molina, A.; Poesen, J.; Deckers, J. & Kubik, P. (2007). Restoring dense vegetation can slow mountain erosion to near natural benchmark levels. *Geology*, Vol. 35, pp. 303-306

Vandekerckhove, L.; Muys, B.; Poesen, J.; De Weerdt, B. & Coppe´, N. (2001). A method for dendrochronological assessment of medium-term gully erosion rates. *Catena*, Vol. 45, pp. 123-161

Viles, H. (1988). *Biogeomorphology.* Basil Blackwell, Oxford

Volsamber, B. & Veen, A. (1985). Digging by badgers and rabbits on some wooded slopes in Belgium. *Earth Surface Processes and Landforms*, Vol.10, pp. 79-82

Wainwright, J., Millington, J.D.A. (2010). Mind, the gap in landscape evolution modelling. *Earth Surface Processes and Landforms*, Vol.35, pp. 842-855

Wainwright, J., Parsons, A.J. (2010). Thornes, J.B. 1985: The ecology of erosion. Classics in physical geography revisited. *Progress in Physical Geography*, Vol.34, pp. 399-408

Wu, T.H.; McKinnell, W.P. & Swanston, D.N. (1979). Strength of tree roots and landslides on Prince of Wales Island, Alaska. *Canadian Geotechnical Journal*, Vol. 16, pp. 19-33

2

Geomorphological Instability Triggered by Heavy Rainfall: Examples in the Abruzzi Region (Central Italy)

Enrico Miccadei, Tommaso Piacentini,
Francesca Daverio and Rosamaria Di Michele
Dipartimento di Ingegneria e Geologia,
Università degli Studi "G. d'Annunzio" di Chieti-Pescara, Chieti scalo (CH),
Italia

1. Introduction

Heavy rainfall is one of the most important triggering causes of landslides - particularly in Mediterranean areas, that are characterised by moderate to low annual precipitation and, occasionally, by high precipitation intensity (up to >200mm/day)-. In agricultural or poorly vegetated hilly landscapes - particularly when characterised by clayey lithologies - heavy rainfall triggers very rapid geomorphological processes, such as floods, soil erosion (rill, gullies) and landslides (rapid earthflows) inducing strong erosion rates on the hilly landscape, sediment transport and sedimentation along the alluvial plains and at the mouths of rivers.

The distribution of geomorphological processes and landforms triggered by these events is variable and controlled by several geological, geomorphological, meteorological and land-use factors. In this work, we analyse the landforms triggered by heavy rainfall in three case studies from the Abruzzi region in Central Italy.

Over the last ten years, the Abruzzi region was affected by several heavy rainfall events. Three of them have had daily rainfall > 100 mm and >200 mm over few days: 1) on 23-25 January 2003 (in the whole region), 2) on 6-7 October 2007 (in a small part of the hilly and coastal Teramo area), and 3) on 1-2 March 2011 (in the hilly and coastal Teramo and Pescara area). These events have triggered different types of geomorphological instability: landslides, soil erosion and flooding. The distribution and types of instabilities and landforms is different in the three cases.

The 2003, 2007 and 2011, heavy rainfall events were analysed with regard to their meteorological aspects, and geological and geomorphological features, highlighting both common and distinct geomorphological effects on the landscape.

The meteorological aspects were studied by processing a >40 pluviometric station database. The data processing enabled the analysis and comparison of hourly rainfall intensity, cumulative rainfall, daily rainfall, monthly rainfall and previous monthly rainfall.

Geomorphological effects of heavy rainfall were analysed through a field surveys, aerial photo analysis and inventories and technical reports, mapping the distribution of the landslides, soil erosion and flooding.

This work allowed us to highlight that these types of methods, investigations and data are basic in applied studies for the stabilisation and management of slopes and minor or major drainage basins, and for general land management. Only a high level of knowledge of geomorphological instability, connected to drainage, geological-geomorphological and morphostructural features and to meteorological events - particularly when joined to geotechnical data - allows effective stabilisation and management plans.

Finally, these types of studies are basic and complementary to recent methods of the investigation and mapping of land sensitivity to such geomorphological processes as landslides, soil erosion and desertification, etc. They allow us to define the future scenarios - which sustainable land planning and management should be based on - by taking into account the specific destination of different areas and contributing to the identification of proper sites for quarries, dumps and purification plants, or else proper areas for industry, urban expansion, thereby generally supporting the process of creating an urban plan.

2. Study area

The Abruzzi region is located in the central eastern part of the Italian peninsula along the Central Apennines (Fig. 1). The regional physiographic and morphostructural setting of Abruzzo is defined by three main orographic and morphostructural domains: the Apennine Chain, Piedmont area and the Coastal Plain (D'Alessandro et al. 2003b).

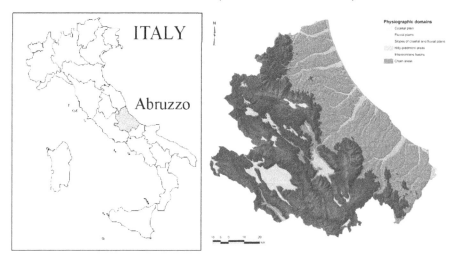

Fig. 1. Location map (a) and main physiographic domains of the Abruzzo region (b).

The hydrography of the region is characterised by three main types of rivers and hydrographic basins, mostly perpendicular to the coast: 1) rivers rising from the inner part of the chain and cutting it transversally, flowing through the piedmont area to the coast; 2) rivers rising from the front of the chain, incising the piedmont down to the coast; 3) rivers rising within the piedmont area and rapidly reaching the coast. A fourth, secondary type - but very important in heavy rainfall events - is given by small catchments flowing on the coastal slopes directly to the coastal plain.

The relief of the Apennine Chain is made up of carbonate ridges (NW-SE, NNW-SSE, N-S) separated by parallel valleys carved in terrigenous foredeep deposits or filled up with continental ones and by wide intermontane basins partially filled with Quaternary continental deposits. To the east, the relief abruptly slopes down into the piedmont area, where a hilly landscape is carved by cataclinal valleys (SW-NE) on arenaceous-pelitic thrusted and faulted successions and on a gently NE-dipping homocline of clay, sand and conglomerate deposits. Along the valleys and close to the coast, alluvial plains with fluvial and alluvial fan deposits join a narrow coastal plain.

The lithologies of the Abruzzo area are made up of different units, mostly of sedimentary origin. In the recent official Geological map of Italy (CARG Project, Geological Survey of Italy, ISPRA, 2011) the lithological units are referable to pre-orogenic units (mostly marine Meso-Cenozoic carbonate rocks), syn-orogenic units (mostly Neogene arenaceous and pelitic rocks), and post-orogenic units (marine Plio-Pleistocene clay-sand-conglomerate rocks and Quaternary clastic continental deposits). These can be grouped in a limited number of units (Fig. 2). These units are mantled, particularly in piedmont slopes and valleys, by eluvial and colluvial, cover up to several meters thick.

The structural setting of the chain area is defined by thrust ridges and faulted homocline ridges separated by tectonic valleys and basins. Main regional fault systems affect the chain area: Mio-Pliocene NW-SE to N-S low angle thrust faults, Pliocene NW-SE to NNW-SSE high angle normal faults, Quaternary (in some cases still active) NW-SE and SW-NE high angle normal faults. The piedmont area is defined in the inner part by thrust reliefs which are affected by regional NNW-SSE Pliocene thrusts and by minor high angle normal faults and, in the outer part, by a wide homocline slightly NE-dipping which is affected by minor high angle normal faults.

The geomorphological processes affecting the whole Abruzzo region are mainly fluvial slope processes and mass wasting. In the coastal areas, marine and aeolian processes are also very important, while in mountain areas karst landforms are present and the landform remnants of ancient Pleistocene glacial processes are preserved. These processes are frequently activated by the heavy rainfall events that affect the region. Fluvial processes affect the main rivers, alternating between channel incisions and flooding. The slope processes that are due to running water mostly affect the clayey and arenaceous-pelitic hills of piedmont and the coastal areas, generating outstanding landforms such as badlands (or "calanchi") and minor landforms, such as rills, gullies and mudflows (which are very common all over the hill slopes).

Mass wasting processes have induced the formation of a huge number of landslides and mass movement in the Abruzzi region, mostly affecting the hilly piedmont area as well as the chain area and - locally - the coastal one (Fig. 2; D'Alessandro et al., 2003a, 2007).

This geological and geomorphological setting is the result of a complex geological and morphostructural evolution due to the Neogene compressional deformation of different Meso-Cenozoic paleogeographic domains and Neogene foredeep domains (that formed a NE-verging thrust belt) and to Quaternary extensional tectonics and uplift (that formed

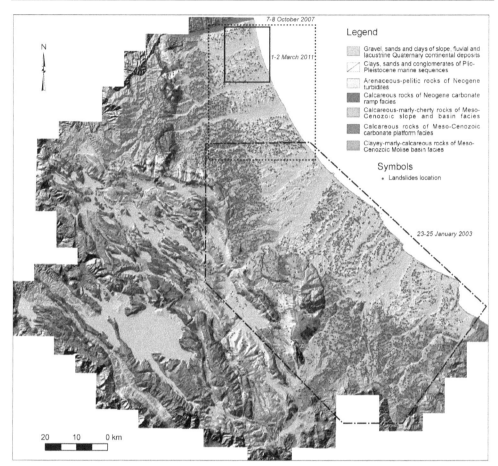

Fig. 2. Geological scheme of the Abruzzo Region (modified from Vezzani and Ghisetti, 1998; ISPRA, 2011); the red dots indicate the location of landslides (modified from Progetto IFFI, D'Alessandro et al., 2007). The black boxes indicate the approximate distribution of heavy rainfall events and related geomorphological effects: a dash-dot line for the 2003 event; a continuous line for the 2007 event; a dashed line for the 2011 event.

faulted homocline ridges, tectonic basins and homocline reliefs) (Cipollari et al., 1997; Lavecchia et al., 2004; Parotto and Praturlon., 2004; Patacca and Scandone, 2007; Cosentino et al., 2010). In the chain as a whole, the morphogenetic processes began during the last phases of thrust belt emplacement in the Early Pliocene. However, the most important morphogenetic impulses are due to the regional uplift processes and the development of extensional tectonics. After the Apennine area emerged, the activity of morphosculptural processes in the continental environment began in a variable relationship with morphostructural processes. This process, together with the morphostructural action of the extensional tectonics, has controlled the geomorphological evolution of the ridge, basin and valley of the chain and in the piedmont relief of the major fluvial valleys systems and the coastal plain. These processes also induced the mantling of slopes and valleys with a cover of Quaternary continental deposits (slope, fluvial, lacustrine and coastal deposits) (Demangeot, 1965; Bigi et al., 1996; Miccadei et al., 1999; D'Agostino et al., 2001; D'Alessandro et al., 2003b; Pizzi, 2003; Farabollini et al., 2004; Miccadei et al., 2004; Ascione et al., 2008; D'Alessandro et al., 2008).

3. Methods

This work is based on the analysis of the meteorological aspects and geomorphological effects of heavy rainfall occurring during the three events affecting the piedmont and coastal area of the Abruzzi region: 1) on 23-25 January 2003 (in the whole region), 2) on 6-7 October 2007 (in a small part of the hilly and coastal Teramo area), and 3) on 1-2 March 2011 (in the hilly and coastal Teramo and Pescara area). The analysis was performed by means of the statistical processing of precipitation data and by means of field surveys, aerial photo analysis and inventories and technical reports.

The meteorological aspects were studied processing a >40 pluviometric station database provided by Servizio Idrografico e Mareografico (Direzione Protezione Civile e Ambiente, Regione Abruzzo), including daily and monthly historical data (30-70 years) and 5-15 min pluviometric registrations for at least six days around the main events. The data processing enabled the analysis and comparison of hourly rainfall intensity, event cumulative rainfall, daily rainfall, monthly rainfall and previous monthly rainfall.

The geomorphological effects of these heavy rainfall events were analysed through field surveys, aerial photo analysis, and inventories and technical reports, and they enabled the mapping of landslides, soil erosion and flooding. The percentage and areal distribution of these effects was also analysed for the different events and so also concerned the affected lithologies, providing a contribution for the definition of the controlling factors' role.

4. Heavy rainfall events

4.1 2003 event (23-25 January)

The 2003 heavy rainfall event affected almost the entire region - but mostly the central and south-eastern part (Chieti and Pescara province and part of L'Aquila and Teramo) - for almost 72 h.

The monthly rainfall analysis shows January 2003 values very much higher than the average historical values, ranging from 50 mm to 150 mm (Fig. 3a), which in some cases is up to 3 times. Several values are >250 mm and some are > 300 mm, up to a maximum recorded value of 388 mm (Salle station; Fig. 3a). These values are close or in some cases higher than the previous historical January maximum values (Fig. 3a). Moreover, the rainfall occurred after a December which had already been very rainy, with a two month precipitation value up to 60% or even 80% of the average annual precipitation (D'ALESSANDRO et al., 2004).

The daily rainfall is high but less than it was for the 2007 and 2011 events. The higher values - up to > 120 mm - are recorded along the front of the chain, in the Maiella area (Salle, Popoli; Fig. 4a).

Hourly rainfall is not as critical as the daily, monthly and event rainfall are. The values do not exceed 9-10 mm/h (Fig. 5a). With regard to the event rainfall, the values are critical, reaching - and in some cases exceeding - 200 mm in three days (Fig. 5a).

In summary, the 2003 event can be considered to be of moderate to high intensity (~10-15 mm/hr, 80-120 mm/day) but very long (72 h), with a high cumulative rainfall (up to >200 mm) which occurred after two months of high rainfall (up to 80% of the average annual rainfall).

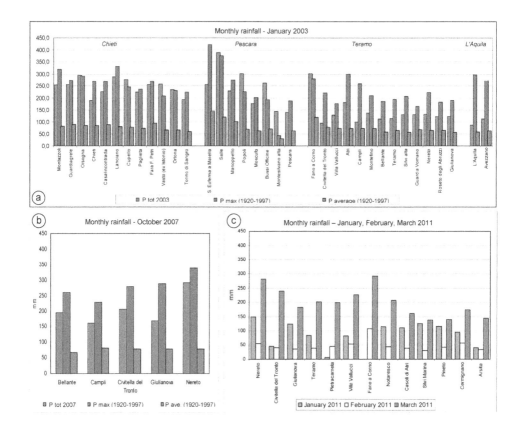

Fig. 3. Monthly rainfall: a) January rainfall histogram for the Abruzzi area; from left to right, data from the pluviometric stations of the Chieti, Pescara, Teramo and L'Aquila provinces. For each station, the total January 2003 rainfall, previous historical maximum January rainfall (1920-1997) and previous average January rainfall (1920-1997) are compared (modified form D'Alessandro et al., 2004); b) October monthly rainfall in the Teramo province's hilly and coastal areas. Comparison of the total October 2007 rainfall, the previous historical maximum October rainfall (1920-2006) and the previous average January rainfall (1920-2006); c) Monthly and daily rainfall; d) 2011 January, February and March rainfall.

4.2 2007 event (6-7 October)

The 2007 heavy rainfall event affected a local area in the northern Abruzzo (hilly and coastal Tortoreto area within the Teramo province, between the Salinello and Vibrata rivers; Fig. 2) for a short time (14-16 h).

The monthly rainfall was higher than the average historical values, ranging from 60 mm to 80 mm (Fig. 3b). The values are >150 mm and some are > 200 mm, up to the maximum recorded value of 291 mm (Nereto station; Fig. 3b). These values are close to the previous historical October maximum values (Fig. 3b). This event occurred after a relatively dry summer period.

The daily rainfall in this case was very intense, with values ~100 mm in the hilly area close to the coast, up to a maximum of 205 mm (Nereto station; Fig. 4b), even when of a short event duration. Along the coast, the recorded daily precipitation is around 60-80 mm (Fig. 4b).

The hourly rainfall was very high, with values from 10 mm/h in the coastal area to 40 mm/h in the hilly area (Nereto station; Fig. 5b).

Taking into account the short duration of the event (14-16 h), the cumulative rainfall during this time interval was also very high, with values from 60-80 mm in the coastal area to 220 mm in the hilly area (Nereto station; Fig. 5b). At the Nereto station, the 24h precipitation recurrence interval is estimated to be between 1000 to 5000 years.

In summary, the 2007 event is an extreme event (for the Mediterranean environment), with a high intensity (10-40 mm/hr, up to >200 mm/day), a high cumulative rainfall (up to 220 mm) and occurring after two months of low rainfall.

4.3 2011 event (1-2 March)

The 2011 heavy rainfall event affected a provincial area (hilly and coastal Teramo area between Vomano, Tordino, Salinello and Vibrata rivers; Fig. 2) for a moderately short time (22-26 h).

The monthly rainfall was very high compared to the average March values. The values are between 150 mm and 250 mm, up to the maximum recorded values of 282 mm (Nereto station; Fig. 3c) and 291 mm (Fano a Corno station; Fig. 3c). This event occurred after a moderately humid winter period comparable with the historical average values.

The daily rainfall was very intense - as in the 2007 event - with values of ~100-120 mm/d in the hilly area close to the coast, up to a maximum of 180 mm (Nereto station; Fig. 4c).

The hourly rainfall was again very high, with values around 15-20 mm/h in the coastal area, and up to 35 mm/h in the hilly area (Pineto and Nereto stations; Fig. 5c).

Taking into account the duration of the event (22-26 h), the cumulative rainfall during this time was again very high, with values from 100-130 mm at most of the stations up to a maximum of 211 mm in the hilly area (Nereto station; Fig. 5c). Also, in this case, at the Nereto station the 24h precipitation recurrence interval is estimated to be between 1000 to 5000 years.

In summary, the 2011 event is an extreme one (for the Mediterranean environment), although intermediate in respect of the previous ones: it affected a provincial area for a moderately short duration (22-26 h) with a high intensity (15-35 mm/h, up to >180 mm/d), high cumulative rainfall (up to 211 mm) and occurring after two months of moderate rainfall.

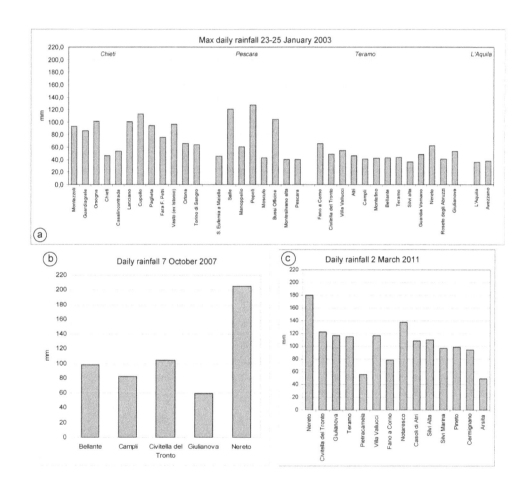

Fig. 4. Maximum daily rainfall occurred: a) on 23-25 January 2003; b) on 7 October 2007; c) on 2 March 2011.

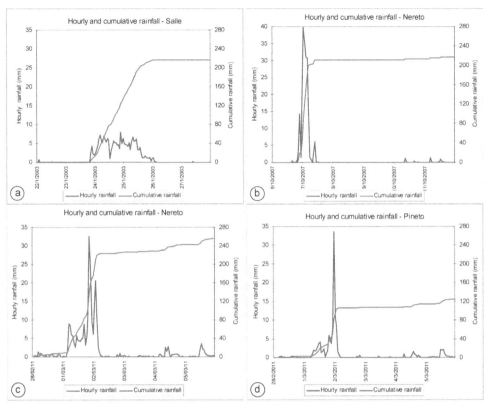

Fig. 5. The hourly and cumulative rainfall occurring during the heavy rainfall events: a) on 23-25 January 2003 at the Salle (PE) station; b) on 7 October 2007 at the Nereto (TE) station; c) on 2 March 2011 at the Nereto (TE) station; d) on 2 March 2011 at the Pineto (TE) station.

5. Geomorphological effects and landforms triggered by heavy rainfall events

5.1 2003 event (23-25 January)

At the end of January, according to the meteorological aspects of the event, outcropping lithologies and surface eluvial and colluvial covers were already very humid and - in almost water saturation conditions - had a strong susceptibility to slope instability. In these conditions, the occurrence of prolonged and intense precipitation on 23-25 January 2003 induced heavy flooding within the main alluvial plains (the rivers Sinello, Sangro, Trigno, Foro and Alento) and triggered ~1300 landslides ranging from the small to the very wide (mostly rapid earth flows and debris flows, secondary rock falls and rotational/translational sliding). The type and distribution of landslides were strictly controlled by the lithology and morphostructural setting and by the poorly vegetated landscape due to the winter season (Fig. 6; D'Alessandro et al., 2004).

The landslides on Quaternary continental deposits (7%) were mostly small flows and slides, located along the scarp edge of fluvial terraces or affecting the talus slopes in the mountain area. The landslides affecting the cuesta and mesa reliefs on the sands and conglomerates of Plio-Pleistocene marine succession of the Abruzzi piedmont area (17%) were distinct type of landslides, on high gradient slopes or else along structural scarps: rapid earth flows affecting surface colluvial cover triggered by heavy surface runoff; falls and topples affecting the edge of structural scarps on sandstones and conglomerates; rotational and translational sliding, which was less frequent but developed for a long time after the event due to deep water infiltration in the permeable conglomerates and sandstones laying on impermeable clays. The landslides on the hilly slopes and cuesta and mesa slopes affecting clays of Plio-Pleistocene marine succession of the piedmont area (32%) were mostly earth flows, from the small to the very wide (Fig. 7a). The landslides on the arenaceous-pelitic and marly rocks of the Neogene syn-orogenic foredeep succession of the piedmont area (15%) consisted of mostly rapid surface flows and sliding (Fig. 7b), affecting the eluvial and colluvial cover, particularly where it is more clay-rich. The landslides on the structural or fault slopes on the marine Meso-Cenozoic carbonate rocks of the chain area ridges (2%) were mainly rockfalls of single blocks or else volumes of fractured rock, which affected several mountain roads (Fig. 7c). The landslides on the slopes and isolated reliefs on the clays and fractured shales ("Argille varicolori" Auctt.) of the Meso-Cenozoic Molise basin succession outcropping in the SE Abruzzo area (27%) were mostly flows and complex landslides occurring on all the slopes with a low gradient, due to the bad geotechnical properties of these rocks.

Finally, floods occurred along the fluvial plain of most of the main rivers (Trigno, Sinello, Sangro, Foro, Alento; Fig. 7d).

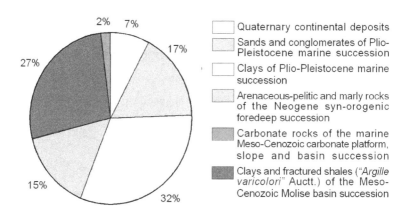

Fig. 6. Percentage distribution (by lithology) of landslides triggered by the 2003 heavy rainfall event (modified from D'Alessandro et al., 2004).

Fig. 7. Examples of landslides and flooding triggered by the 2003 heavy rainfall event (modified from D'Alessandro et al., 2004): a) Atessa, rapid earth flow on clayey and sandy deposits; b) Fresagrandinaria; slides on pelitic-arenaceous deposits; c) Popoli, rockfall on carbonate rocks; d) F. Salinello, flooding.

5.2 2007 event (6-7 October)

The event affected a very poorly vegetated landscape of the hilly coastal area, particularly the agricultural areas (arable land, vineyards and olive groves) due to ploughing on erodible rocks (marine Plio-Pleistocene clays, sands and conglomerates) and clayey eluvial and colluvial cover. This provided a strong propensity towards soil erosion.

These features induced heavy soil erosion processes on slopes (sheet, rill, and gully erosion), rapid mud flows at the base of slopes or minor drainage basins, and flooding within the main river and coastal plains, mostly at the outlet of minor tributary catchments. The distribution of landforms is controlled by the orography of the basins (slope and aspect) and by land use, particularly with regard to vegetation cover.

The landform analysis shows that hilly catchments and slopes, during this event, were affected by gully erosion (31% of the affected area, Fig. 8) and sheet and rill erosion (35% of the affected area, Fig. 8) for an extent up to >50% of the total area. The low gradient slopes and ridges with the vineyards and olive groves - or in general those not ploughed - were mostly affected by low sheet-rill erosion. The high gradient slopes - particularly where ploughed in down slope direction - are incised by gullies up to ~1m deep and 3-4 m spaced.

Along the channel of the main catchments, channel incision occurred. Also very common were mud flows (6% of the affected area, Fig. 8) and flooding (29%, Fig. 8) at the bases of the slopes, along the channel of small catchments or at the junction between small catchments and the flood plain or coastal plain. Mud flows are the result of heavy soil erosion on the slopes, inducing the mobilisation of huge sediment volumes of silts and clays from the eluvial and colluvial cover. Extensive mud and water flooding affected both coastal plains, coming more from the small catchments of the coastal slopes and river plain slopes, more than from the main rivers.

As such, the landform distribution seems to be controlled by the structural geomorphological features of the radial drainage pattern, incorporating the coastal streams and tributary streams of the main valleys of F. Salinello and T. Vibrata.

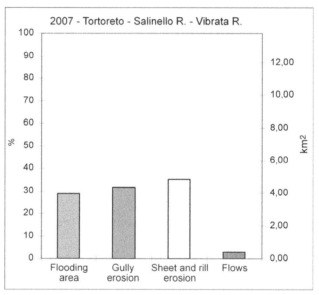

Fig. 8. Percentage and surface distribution of the geomorphological instabilities triggered by the 2007 heavy rainfall event in the Tortoreto hilly and coastal area between the lower T. Vibrata valley and lower F. Salinello valley.

5.3 2011 event (1-2 March)

This event occurred after two months of moderate rainfall on moderately humid clay-sands-conglomerate rocks and clayey eluvial and colluvial cover. In respect of the 2007 event - which partially affected the same area - the 2011 one occurred on a moderately vegetated landscape with agricultural areas (arable land, vineyard and olive groves) in an initial crop growth stage and grass development, less susceptible to soil erosion. Crop and grass cover promoted surface runoff directly into the main rivers, protecting the slopes from soil erosion.

The geomorphological analysis performed after the event outlined the landforms due to sheet and rill erosion, gully erosion, mud flows and flooding, as with the 2007 event but with a different proportion and a minor total extent. Also, in this case, the distribution is controlled by the catchments' orography and land use, with especial reference to crop and grass cover.

Soil erosion occurred on a small extent (9-17% of the area affected by instabilities; Fig. 9) as well as rapid mud flows (4-7%, Fig. 9) due to the vegetation cover. Low gradient slopes were affected by sheet and rill erosion. High gradient slopes were affected by mud flows, at the base of the slopes (Fig. 10a), and by gullies up to 0.6 m deep and sparse in respect of the 2007 event. Gullies occurred on the channels of tributary catchments or on slope undulations, in some cases enlarging natural or man-made notches or else simply notches due to agricultural work (Fig. 10b). Along the main rivers, channel incision or lateral erosion occurred (Fig. 10c,d), inducing severe damage to valley roads and bridges (Fig. 10e,f).

Heavy mud and water flooding is the prevailing effect of this event (75% of the area affected by instabilities; Fig. 9) and it affected almost seamlessly the coastal plains between the Tronto river and the Piomba river as well as the river plains of the Salinello, Tordino, Vomano, and Calvano rivers. Coastal plains flooding came from the small coastal slope catchments and partly from the main rivers, while river plains flooding mostly came from the main rivers' overbank flow and - secondarily - from the tributary catchments' flow. Extensive overbank flooding along main the rivers induced the formation of wide and long crevasse splays on the floodplain (Fig. 10c,g). As with the 2007 event, but to a lesser degree, the junction between slopes, small catchments and plains was affected by rapid mud flows.

Also, in this case, and finally, runoff and flooding were controlled by the structural geomorphological features of the radial and trellis drainage pattern of the Teramo hilly area.

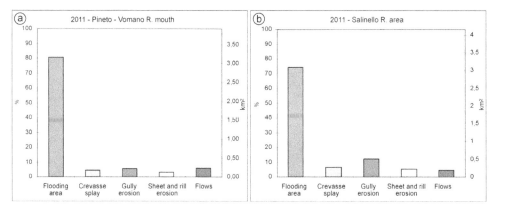

Fig. 9. Percentage and surface distribution of landforms triggered by the 2011 heavy rainfall event: a) The Pineto coastal and hilly area between F. so Foggetta and F. Vomano; b) The Lower F. Salinello valley and the hilly and coastal slopes of the Tortoreto area.

Fig. 10. Landform triggered by the 2011 heavy rainfall event: a) Mosciano S. Angelo, rapid earthflows; b) Pineto, gullies on the coastal slope; c) F. Salinello, crevasse splay and fluvial erosion scarps; d) F. Salinello, fluvial erosion scarp affecting the valley road; e) F. Vomano, main river flooding affecting main roads; f) F. Salinello, damage to a bridge; g) F. Salinello, flooding and crevasse splays along the main alluvial plain.

6. Conclusion

This work deals with the distribution of flooding and mass movements triggered by heavy rainfall events, analysing the effects of three events which occurred over the last ten years in the Abruzzo region (Central Italy). The analysis is carried out with regard to the geographical extent of the events (regional-local), its meteorological and pluviometric features (monthly, daily, hourly and cumulative rainfalls), the lithological and morphostructural setting, land use (also concerning the vegetation state and agricultural maintenance of cropland, olives and vineyards). The three events that are analysed all had different features (Tab. 1), concerning:

- geographical extent (2003 regional; 2007 local; 2011 intermediate);
- lithological setting;
- duration (2003 ~3 days; 2007 <1 day; 2011 ~1 day);
- season of occurrence (2003 winter; 2007 autumn; 2011 winter end);
- previous humidity conditions (2003 very humid; 2007 very dry; 2011 moderately humid).

This variable conditions, taking into account also the general geomorphological setting and landslide distribution of the Abruzzo region, induced the trigger of different geomorphological instabilities (landslides, mass movements), concerning type and areal distribution, as summarised by Table 2.

Event	Date	Extent	Season	Durat. (hours)	Ih_{max} (mm/h)	Pd_{max} (mm)	Pc_{tot} (mm)	Pm_{tot} (mm)	P previous
2003	23-25 gen	regional	winter	~72	10-17	40-130	80-230	120-380	elevate
2007	6-7 ott	local	autumn	14-16	10-40	60-205	60-220	200-300	scarce
2011	1-2 mar	intermediate	winter end	22-26	15-35	60-180	120-211	150-300	moderate

Table 1. Main meteorological characteristics of the three heavy rainfall events studied in this work. Legend: Ih_{max} - maximum hourly rainfall intensity during the event; Pd_{max} - maximum daily rainfall during the event; Pc_{tot} - cumulative rainfall during the event; Pm_{tot} - total rainfall during the event's month; P previous - rainfall in the month before the event.

Event	Landslides	Flooding	Gullies	Rills and sheet erosions	Crevasse splays
2003*	>1300 landslides	Alento, Foro, Sangro, Sinello, Trigno	n.d.	n.d.	n.d.
2007	~ 0,6 km²(6%) flows	~ 3,8 km² (29%)	~ 4,0 km² (31%)	~ 4,5 km² (35%)	-
2011	~ 0,5 km² (6%) flows	~ 6,5 km² (75%)	~ 0,8 km² (9%)	~ 0,4 km² (5%)	~ 0,5 km² (6%)

Table 2. Geomorphological instability and landforms triggered by the three heavy rainfall events studied in this work. *Data for the 2003 event are from D'Alessandro et al., 2004.

This work allowed us to highlight that these types of methods, investigations and data are basic to applied geomorphological studies for the stabilisation and management of slopes and minor or major drainage basins as well as for general land management. Only a high level of knowledge of geomorphological instability - connected to drainage - geological-geomorphological and morphostructural features and meteorological events - particularly when joined to geotechnical data - allows for effective stabilisation and management plans.

Finally, these types of studies are basic and complementary to recent methods for the investigation and mapping of land sensitivity to geomorphological processes, such as landslides, soil erosion and desertification, etc. (Mitasova *et al.*, 1996; Fluegel *et al.*, 2003; Grimm *et al.*, 2003; Agnesi *et al.*, 2006; Ciccacci *et al.*, 2006; ISPRA, 2006; Marker *et al.*, 2008; and references therein), and they allow us to define future scenarios - which sustainable land planning and management should be based on - taking into account the specific destination of different areas and contributing to the identification of proper sites for quarrying, dumping and purification plant, etc., or else the proper areas for industries, urban expansion or supporting in general the process of creating an urban plan.

7. Acknowledgements

The authors wish to thank the Servizio Idrografico e Mareografico, Direzione Protezione Civile e Ambiente, Regione Abruzzo (Dr. Monica di Michele) for providing pluviometric data of the events and the historical pluviometric data analysed in this work as well as for their useful discussions and suggestions concerning the data processing. The authors wish also to thank the Struttura Speciale di Supporto Sistema Informativo Regionale of Abruzzo Region (http://www.regione.abruzzo.it/xcartografia/), for providing the topographic data and aerial photos used for the geomorphological investigations.

8. References

Agnesi V., Cappadonia C., Conoscenti C., Di Maggio C., Màrker M., Rotigliano E. (2006). Valutazione dell'erosione del suolo nel bacino del Fiume San Leonardo (Sicilia centro-occidentale, Italia). In: Atti del Convegno conclusivo del Progetto di Rilevante Interesse Nazionale "Erosione idrica in ambiente mediterraneo: valutazione diretta e indiretta in aree sperimentali e bacini idrografici", G. Rodolfi ed., Brigati, Genova, pp. 13-27.

Ascione A., Cinque A., Miccadei E., Villani F. (2008). The Plio-Quaternary uplift of the Apennine Chain: new data from the analysis of topography and river valleys in Central Italy. Geomorphology, 102, pp. 105-118.

Bigi S., Calamita F., Cello G., Centamore E., Deiana G., Paltrinieri W., Ridolfi M. (1996). Evoluzione messiniano-pliocenica del sistema catena-avanfossa dell'area marchigiano-abruzzese esterna. Studi Geol. Camerti. vol. spec., 1995/1, pp. 29-37.

Ciccacci S., Del Monte M., Fredi P., Lupia Palmieri E. (2006). Sviluppo di metodi per la valutazione quantitativa dell'intensità della denudazione e della relativa pericolosità in bacini idrografici italiani. In: Atti del Convegno conclusivo del Progetto di Rilevante Interesse Nazionale "Erosione idrica in ambiente mediterraneo: valutazione diretta e indiretta in aree sperimentali e bacini idrografici", G. Rodolfi ed., Brigati, Genova, pp. 125-144.

Cipollari P., Cosentino D., Parotto M. (1997). Modello cinematico-strutturale dell'Appennino centrale. In: Cello G., Deiana G., Pierantoni P.P. "Geodinamica e tettonica attiva del sistema Tirreno-Appennino". Studi Geol. Camerti, vol. spec. 1995/2, pp. 135-143.

Cosentino D., Cipollari P., Marsili P., Scrocca D. (2010). Geology of the central Apennines: a regional review. In: Beltrando M., Peccerillo A., Mattei M., Conticelli S., Doglioni C. Eds. The Geology of Italy, Journal of the Virtual Explorer, Electronic Edition 36: paper 11.

D'Agostino, N., Jackson, J.A., Dramis, F., Funiciello, R. (2001). Interactions between mantle upwelling, drainage evolution and active normal faulting: an example from central Apennines (Italy). Geophysical Journal International 141, pp. 475-497.

D'Alessandro L., Berti D., Buccolini M., Miccadei E., Piacentini T. and Urbani A. (2003a) - Relationships between the geological-structural framework and landslide types in Abruzzi (Central Apennine). In: "Atti 1° Congresso Nazionale AIGA", Chieti, 19-20 Febbraio 2003", pp. 255-275.

D'Alessandro L., Del Sordo L., Buccolini M., Miccadei E., Piacentini T., Urbani A. (2007) – Regione Abruzzo (Cap.18). In: Rapporto sulle frane in Italia. Il Progetto IFFI. Risultati, elaborazioni, e rapporti regionali. Rapporti APAT 78/2007, pp. 464-497.

D'Alessandro L., Genevois R., Paron P., Ricci F. (2004). Type and distribution of rainfall-triggered landslides in the Abruzzi region (central Italy) after the January 2003 meteoric event. 32nd International Geological Congress: "From the Mediterranean Area toward a Global Geological Renaissance. Geology, Natural Hazards and Cultural Heritage". Firenze 20-28 August 2004.

D'Alessandro L., Miccadei E., Piacentini T. (2003b). Morphostructural elements of central–eastern Abruzzi: contributions to the study of the role of tectonics on the morphogenesis of the Apennine chain. In: "Uplift and erosion: driving processes and resulting landforms", International workshop, Siena, September 20 - 21, 2001. Quaternary International, 101-102C, pp. 115-124, Elsevier Science Ltd and INQUA, Oxford U.K.

D'Alessandro L., Miccadei E., Piacentini T. (2008). Morphotectonic study of the lower Sangro River valley (Abruzzi, Central Italy). In: P. G. Silva, F.A. Audemard and A. E. Mather Eds. "Impact of active tectonics and uplift on fluvial landscapes and drainage development". Geomorphology 102, pp.145–158, Elsevier B.V., doi: 10.1016/j.geomorph.2007.06.019.

Demangeot J. (1965). Géomorphologie des Abruzzes adriatiques. Mem. Et Documents, Centre Rech. et Docum. Cartograf. Et Géogr, n.h.s., Edit. C.N.R.S. Paris.

Farabollini P., Miccadei E., Pambianchi G, Piacentini T. (2004). Drainage basins analysis (forms and deposits) and Quaternary tectonics in the Adriatic slope of Central Apennines (Central Italy). 32nd International Geological Congress: "From the Mediterranean Area toward a Global Geological Renaissance. Geology, Natural Hazards and Cultural Heritage". Firenze 20-28 August 2004.

Fluegel W.A., Maerker M. Moretti S., Rodolfi G. (2003). Integrating GIS, Remote Sensing, Ground Trouthing and Modelling Approaches for Regional Erosion Classification of semiarid catchments in South Africa and Swaziland. Hydrological Processes, 17, pp. 917-928.

Grimm M., Jones R.J.A., Rusco E., Montanarella L. (2003). Soil erosion risk in Italy: a revised USLE approach. EUR 20677 EN, European Soil Boreau Research Report, vol. 11.

Office for Official Publications of the European Communities Luxembourg, Luxemburg, pp. 28.

URL: http://eusoils.jrc.it/ESDB_Archive/eusoils_docs/esb_rr/n11_ita_er06.pdf

ISPRA Servizio geologico d'Italia (2011). Cartografia Geologica d'Italia alla scala 1:50.000 (Progetto CARG).

URL: http://www.apat.gov.it/site/it-IT/Servizi_per_l%27Ambiente/Carte_geolog iche/

ISPRA (2006). La lotta alla desertificazione in Italia: stato dell'arte e linee guida per la redazione di proposte progettuali di azioni locali. Manuali e linee guida - 41/2006. ISBN 88-448-0213-9.

URL: http://www.apat.gov.it/site/it-IT/APAT /Pubblicazioni /Manuali_ e_linee _guida/Documento /manuale_2006_41.html

Lavecchia G., Boncio N., Creati N. and Brozzetti F. (2004). Stile strutturale e significato sismogenetico del fronte compressivo padano-adriatico: dati e spunti da una revisione critica del profilo Crop 03 integrata con l'analisi di dati sismologici. Boll. Soc. Geol. It., 123, pp. 111-125.

Märker M., Angeli L., Bottai L., Costantini R., Ferrari R., Innocenti L., Siciliano G. (2008). Assessment of land degradation susceptibility by scenario analysis: a case study in Southern Tuscany, Italy. Geomorphology, 93, pp. 120–129.

Miccadei E., Barberi R., Cavinato G.P. (1999). La geologia quaternaria della Conca di Sulmona (Abruzzo, Italia centrale). Geol. Romana, 34, pp. 58-86.

Miccadei E., Paron P. and Piacentini T. (2004). The SW escarpment of the Montagna del Morrone (Abruzzi, Central Italy). geomorphology of a faulted-generated mountain front. Geografia Fisica e Dinamica Quaternaria, 27, pp. 55-87.

Mitasova H., Hofierka J., Zlocha M., Iverson L.R. (1996). Modeling topographic potential for erosion and deposition using GIS. International Journal of GIS, 10, pp. 629–641.

Parotto M. Praturlon A., (2004), 'The southern Apennine arc. In: Geology of Italy', Special Volume of the Italian Geological Society for the IGC 32 Florence, pp. 53-58.

Patacca E., Scandone P. (2007). Geology of the Southern Apennines. Boll. Soc. Geol. It. Spec Issue CROP-04, 7, pp. 75-119.

Pizzi A. (2003). Plio-Quaternary uplift rates in the outer zone of the central Apennines fold-and-thrust belt, Italy. Quaternary International, 101-102C: pp. 229-237.

Vezzani L., Ghisetti F. (1998). Carta Geologica dell'Abruzzo, scala 1:100.000. Regione Abruzzo, set. Urbanistica-Beni Ambientali e Cultura.

3

Using Discrete Debris Accumulations to Help Interpret Upland Glaciation of the Younger Dryas in the British Isles

W. Brian Whalley

Department of Geography, University of Sheffield, Sheffield, United Kingdom

1. Introduction

With no glaciers in the British Isles in the last 10, 000 years or so, the acceptance of the 'Glacial theory' propounded by Agassiz and Charpentier in the Alps in the 1830-40s was late to be accepted in the British Isles (Chorley, et al., 1973). Scientists from the British Isles had travelled to areas with glaciers, yet even popular tourist areas such as the English Lake District area were not considered to have been affected by glaciers until the 1870s (Oldroyd, 2002). In Scotland however, the ideas of James Croll, giving a theoretical reason for changes in climate, were persuasive in an earlier acceptance of glacial interpretations (Oldroyd, 2002). Field mapping by the Geological Survey in Edinburgh helped to displace the 'floating iceberg' and 'diluvial' theories, especially in the explanation of erratics (Rudwick, 2008). Once accepted, the basic tools of mapping the former extent of glaciers, for example by the recognition of moraines, became commonplace. More recently, aerial and satellite imagery have made drumlins and cross-cutting depositional modes important in elucidating the limits of the terrestrial Pleistocene glacial record (Clark et al., 2004). However, and perhaps inevitably, interpretations of the significance of some moraines and their corresponding glaciers has been in debate. Together with chronological methods, detailed mapping of glacial limits and altitudes allow comparisons with climatological models and general climatic interpretations such as the recent interpretations of the Younger Dryas in Scotland summarised by Golledge (2010).

In this paper I take an overview of the problems associated with a variety of features, other than moraines, associated with mapping the glacial limits (and associated climatic conditions) in the Younger Dryas Stadial in the British Isles. It does not aim to be comprehensive in treatment of these features in the British Isles but is concerned with the problems of mapping and interpreting a variety of features. Recognising the genesis is important as it may help to provide evidence for the magnitude-frequency of selected events as well as help to distinguish between a variety of events that may produce similar-looking landforms. Furthermore, as some features seen and mapped may be post-glacial slope failures rather than glacial deposits, their identification and correct interpretation may be useful for mapping slope failures in an area rather than glacial features. First however, it is necessary to identify some terms and meanings that will be used or have been used in mapping the Late Holocene in the uplands of the British Isles.

2. The Younger Dryas (YD) in the British Isles and the dating of events

In the British Isles, the cold period known as the Younger Dryas Stadial (12.8 – 11.5 ka BP) (Muscheler et al., 2008) but also, and more usually, considered to be 11 – 10 ka BP. It is also known as the Loch Lomond Stadial after the large inland loch to the north of Glasgow. The last stage of the Last Glacial Maximum in the British Isles is generally taken to be the Dimlington Stadial (Rose, 1985) 26 – 13 ka BP as part of the Late Devensian Glaciation. As such, the Younger Dryas saw a deterioration (increasingly cold and wet) period in which glaciers advanced (or grew) again. It followed a relatively warm period known in the British Isles as the Windermere Interstadial (= Allerød). The variability of the dating the YD may be related to when the cooling stage started and its severity. That the YD is generally agreed to be a world-wide phenomenon (Ivy Ochs et al., 1999) with glacier advances being seen e.g. in the Colorado Rockies of interior USA (Menounos and Reasoner, 1997) as well as more maritime areas such as the British Isles. It is also suggestive that the timing may not be exactly coeval everywhere but may indicate that responses to climate change may differ, e.g. in latitude, altitude as well as 'continentality' across the islands. These uncertainties need to be taken into consideration when viewing the landforms and processes in this paper. For example, the date of a recessional moraine of a glacier in the Alps may be known to a year but this is highly unusual in moraine sequences as far back as the YD. Even dated trees need to be put in context. Similarly the known maximum advances of glaciers in the Little Ice Age (again variously defined in terms of chronology but generally taken to be 1600 – 1850 CE). Examples from the Alps and Pyrenees as well as elsewhere provide some near present-day analogues that help interpretation in the British Isles.

In upland areas, where the Younger Dryas glaciers may have been small and reacting to subtle variations in a rapidly changing climate, the analysis may require careful mapping and interpretation for specific areas. This is shown by Benn and Lucas in their landsystems approach in NW Scotland (Benn & Lukas, 2006). They use present-day analogues to help their interpretation much in the same way that Hauber et al. (2011) have used Svalbard to provide periglacial analogues for Martian landforms. The use of analogues is used generally for the interpretation of surface features in planetary geology (Farr, 2004). This use is appropriate as periglacial and possibly permafrost features are associated with upland regions of the British Isles in the Younger Dryas. Compilations of processes, mechanisms and chronologies can be found in Ballantyne and Harris (1994) and Gordon and Sutherland (1993) and other overviews have been provided in several other volumes (Boardman, 1987; Gillen, 2003; Gordon, 2006; Gordon & Sutherland, 1993; McKirdy et al., 2007; Wilson, 2010).

A distinction should be made be made between periglacial, that is, around a glacier and permafrost, a thermal condition where the mean annual air temperature is assumed to be <-2°C. So a snowbank is generally assumed to be a periglacial feature but is not glacial, ie with the dynamics of a glacier-ice body. Neither condition is extant in the British isles at the present day and we shall see that may produce interpretational difficulties. The term paraglacial has been used to indicate features that are postglacial and are involved with sediment movement (Ballantyne, 2002; 2007). It was instituted into the literature by Church and Ryder (1972) as, materials that were produced by 'non-glacial processes that are conditioned by glaciation'. Ballantyne (2007) defines it as 'the study of the ways in which glaciated landscapes adjust to nonglacial conditions during and after deglaciation'. However, in many cases this does not help with the interpretation as it may not be at all clear what was glacier or snow or permafrost-related. Further discussion can be found in Slaymaker (2009).

If there are problems in interpreting the significance of moraines this is also true of landscape features where the ice-debris mix is of less certain origin and formative process unclear. For example, the ice-cored moraines investigated by Østrem in Scandinavia (Østrem, 1964) were interpreted by Barsch (1971) as being 'rock glaciers'. This dispute (Østrem, 1971) is still not resolved. There are several reasons for this uncertainty; problems of observation as well as nomenclature and understanding of the geological processes and mechanisms involved and their rates of operation. This is despite advances of glacial theory, sedimentology and dating techniques. Additionally, researchers coming from diverse backgrounds have tended to have different, often divergent, views about the processes operating and therefore the interpretations. Further discussion on this will follow below.

3. Discrete Debris Accumulations and terminology

The term Discrete Debris Accumulations (DDA) has been used to encompass a range of features that can be mapped in the field or from aerial photographs (Whalley, 2009). It is used here as a non-genetic and descriptive term, such that focus can be given to whatever is under study without any preconceived notion of origin or significance. The actual interpretation of these features is, in the British Isles, very much related to the use of analogues. This is especially significant when the presence of ice (of some origin) is considered and so the recognition of ice-debris features and their mechanisms is considered. Although DDAs *may* be paraglacial it could be that some are fossil glacial features and not at all modified by post YD activity. Hence, there needs to be some care in distinguishing between periglacial, proglacial, paraglacial and permafrost in these studies.

This paper is specifically concerned with recognising the process and mechanisms of debris accumulations rather than dating *per se*. In particular, the association of specific features can be associated with climatic conditions. For example, moraines are associated with glaciers and the size (mass balance) of the glaciers. In the British Isles there is an assumed west-east gradient in glacier net balance, such has been found in Scandinavia (Chorlton & Lister, 1971). There are also possibilities of changes in prevailing winter storm tracks that may influence the size of glaciers (Whalley, 2004) that have not yet been investigated for the palaeo-conditions for the British Isles compared with suggestions for northern Scandinavia (Bakke et al., 2005; 2008).

The traditional view of the relationship between glacier extent, mass balance and glacial record is the linear set of boxes in Figure 1. The 'geological record' is usually taken as being manifest in the simplest (or least complicated) debris accumulation associated with glaciers; a moraine. Although the basic idea may hold, interpretation is more complicated for periglacial features such as snowbanks and their rock debris remnants such as 'protalus lobes'. Is such a feature classed as periglacial, proglacial or indeed paraglacial (Slaymaker, 2007; 2009)? This problem will be considered in more detail below.

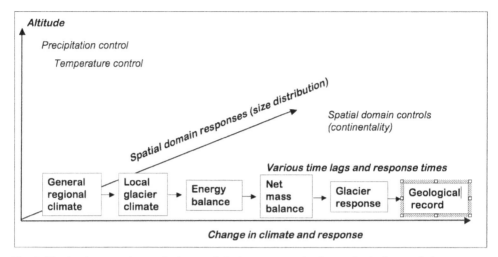

Fig. 1. The basic controls on glaciers and their responses in the geological record; from Whalley (2009) after Andrews (1975) and Meier (1965). The boxed sequence is embedded within domains that ultimately affect the geological record.

4. Mapping Discrete Debris Accumulations

Despite remote sensing techniques, direct field observation is still important when process recognition remains a problem. In this paper it is suggested that care must be used when interpreting landforms, especially when related to their past climatic history. This is especially important where rates of process are assumed and where similar landforms might be produced by different processes ('equifinality' or 'form-convergence').

Weathered rock debris accumulations, whether directly deposited from a glacier or by some creep or flow mechanism, frequently have distinct forms, to which names are given – although the origins may be disputed. Such discrete debris accumulations can be mapped. To interpret these forms, especially to make inferences about environmental conditions, observations have to be placed within the context of imperfect knowledge of behaviour and response to past environmental conditions or events. This paper suggests that caution and more precise glacio-geomorphological investigations are required. Selected examples illustrate such problems from present day analogues and from ice-free areas.

The geological literature has many examples of differing or changed opinions about features – in the widest sense. For example, in the present context, Wilson (2004) now views certain rock glaciers in Donegal (Ireland) as (paraglacial) rock slope failures. This changes the paleo-environmental interpretation from being related to permafrost to one where permafrost (nor glacier) were involved.

A change in climate leading to glacier mass balance change and a glacier leaving an interpretable and dated trace (such as a moraine) is useful in a regional as well as temporal manner. Shakesby and co-authors (Shakesby, 1997; Shakesby & Matthews, 1993; Shakesby et al., 1987) have discussed problem related to protalus ramparts. There are different responses according to glacier size as well as the mode of precipitation input (winter storms, summer monsoon) and the effects of continentality as suggested above. For example, Harrison et al. (1998) suggested that a small glacier existed in the lee of the Exmoor plateau. This was based on their interpretation of a small moraine or protalus rampart at the foot of a small valley head (combe/corrie/cirque). Indeed, the use of the term 'moraine' or 'protalus lobe' may well indicate differences of interpretation. Some of these, perhaps indistinct, features, tend to be problems of interpretation rather than mapping. Certain debris accumulations in the English Lake District (Sissons, 1980) present problems of climatic interpretation, especially when it is not clear what the features represent in terms of debris and ice input. Similarly, Harrison et al. (2008) have discussed features that have generally been called 'rock glaciers' and their environmental significance. More precise matching of formation processes and mechanisms to environmental conditions will help to improve modelling of ice mass extents and volumes and associated climatic environments (Gollege & Hubbard, 2005; Hubbard, 1999).

5. Debris input to glacial systems

To the basic glacial parameters of Figure 1 weathered rock debris needs to be added to the system for there to be traces in the geological record. This complication is rarely considered; not just a morainic marker but to consider where, when and from where the debris addition was made. It may well have a considerable effect on the system as a whole. The total amount may be important. Dead ice preserved at the snout of a glacier long after the debris-free glacier has melted may give hummocky moraine or even a rock glacier form. Further, the debris flux may have an effect upon the ice extent. For example, a glacier in equilibrium that receives a debris input near the snout (as from a large rockfall) could produce a glacier advance as the ablation area is reduced. The timing of debris input (at the start or end of a glacier advance phase for example) may be significant. Some work has been done on this in the British Isles, eg. (Ballantyne & Kirkbride, 1987). Although dating such events via cosmogenic ratio methods is now becoming easier (Ballantyne & Stone, 2004; Ballantyne et al., 1998) care must be taken in the temporal interpretation.

Figure 2 indicates the potential complexity here, again related to altitude, continentality and temporal input variations. Rock debris can be added to a glacial, permafrost or periglacial system. Unknowns include the relative and absolute amounts of ice/water and debris but also the flux changes in time (Nakawo et al., 2000; Whalley et al., 1996). Even 'simple' glacial systems may show this. For example a large rockslide on or near the snout of a glacier may allow the snout to advance but if away from the snout the glacier may be hardly affected In the relatively small glacier in the British Isles however, substantial, but largely unknown, effects may be produced.

Figure 3 illustrates possible scenarios produced by the relative additions of weathered debris quantities to snow/ice bodies. This should not be taken to show that certain features *will* form but that they are possible given the relative components at any time. For example, the time element is not considered as part of the formative process. Some features might be form 'rapidly', others take some time. For the most part process studies do not give a good indication of the time needed to produce a feature. If the debris is lacking then there may be no feature formed at all. However, the diagram does suggest that there is a continuum of features and it does help guide interpretation of what is seen or mapped.

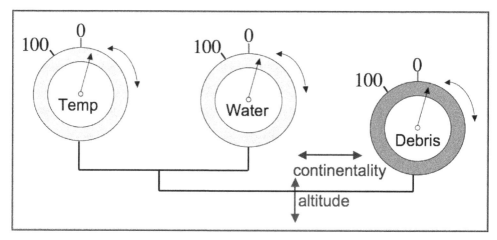

Fig. 2. An illustration of the weathered rock debris constituent needs to be taken into account when considering 'glacial, proglacial, perglacial or permafrost conditions. Not only may the debris addition be sudden (rock avalanche) or slow and continuous (scree formation); after Whalley (2009).

A further formational aspect not shown in Figure 3 are the possible altitude-temperature/precipitation-continentality controls (Figures 1 and 2). Thus, it is by no means clear where the 'best' analogues for the YD in the British Isles should be taken. For example, it was once thought that rock glaciers were only found in 'continental' mountains until examples from Iceland were found. The answer lay in the relative amounts of debris supplied to small glacier systems. Furthermore, Icelandic rock glaciers are found where there is no (or only sporadic) permafrost. Hence, the inverse interpretation; relict rock glacier = former permafrost, needs to be used carefully. This applies in fact to most of the features here classed as DDAs

6. Rock debris production in upland British Isles

As the ideas shown in Figures 2 and 3 depend upon debris input some consideration will now be given to the production of rock debris. At the present day, however, there is very little active rock fall production. There are some active scree slopes (talus), as shown by the lack of vegetation, but these are relatively uncommon.

Rockfalls (of indeterminate size) may be associated with periods of cliff instability related to a number of possible factors. These include glacial unloading and seismic, neotectonic, tremors associated with isostatic readjustment (e.g. Jarman, 2006) large-scale weakening of rock buttresses caused by intense periglacial weathering; permafrost melting or a combination of these factors. Some characteristics of these rock glaciers includes: location within mapped Younger Dryas glacier limits, high and steeply-angled cliffs upslope of the landform and largely unvegetated surfaces. A review of large 'felssturz' (rockfall events) in extra-glacial areas of Austria by (Meissl, 1998) shows how significant such events may be even today. There is compelling evidence (Whalley, 1984; Whalley et al., 1983) that near-glacial conditions were, and are, important in the development of rockfalls and slides.

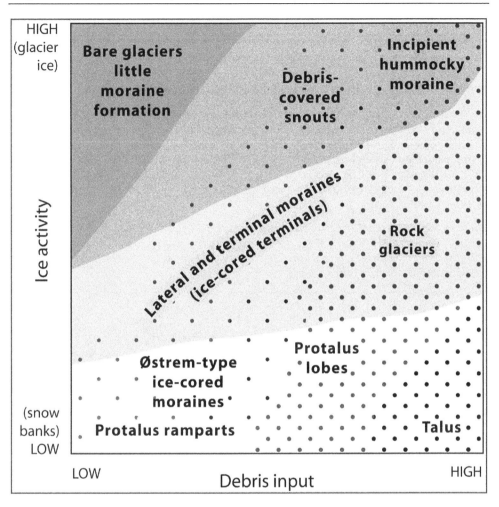

Fig. 3. A schema illustrating the relative proportions (and perhaps fluxes) of snow/ice and rock weathering debris in a 'glacial' geomorphic system. From Whalley (2009).

Permafrost warming post Younger Dryas may also have had a significant part to play as has been suggested for present-day rockfall production and (Davies et al., 2003; Whalley et al., 1996) have shown that large debris accumulations are often associated with Little Ice Age events. It is not yet clear how substantial and variable was the production of debris in the Younger Dryas, although some attempts have been made (Ballantyne & Kirkbride, 1987). More recently, Jarman (2009) and Wilson (2009) have examined rockfalls and slope failures associated with Younger Dryas slope activity and the production of discrete debris accumulations. What does seem to be the case is that fossil rock glaciers and protalus lobes are relatively rare in the British Isles and Ireland compared with many mountainous regions. This may well be a consequence of the lack of weathering or rockfalls from the Caledonide rocks that comprise much of upland Britain (Harrison et al., 2008). It is perhaps not surprising that Norway, similar in a geology of hard old rocks, also seems deficient in rock glaciers and protalus lobes. Unsurprisingly however, present-day scree formation in Norway does seem more active than in Britain because of more severe weathering conditions.

7. Mechanical properties of ice and ice-rock mixtures

Having thus indicated the importance, yet variability of ice/water/debris fluxes within mountain systems in the British Isles in Younger Dryas times we now need to consider what the effects might be upon the mechanical properties of the materials produced. This has been considered in several papers (Whalley, 2009; Whalley & Martin, 1992; Whalley & Azizi, 1994; 2003) and will not be elaborated upon here. Figure 4 illustrates graphically another continuum, of strength or flow of ice according to the content or dispersion of rigid rock particles.

We have to use present-day analogues and known rheological behaviour to interpret past deposits. Unfortunately, even the present day features may be in dispute. This makes interpretation of Younger Dryas DDAs even more problematic, hence even more to interpret past climate from such features.

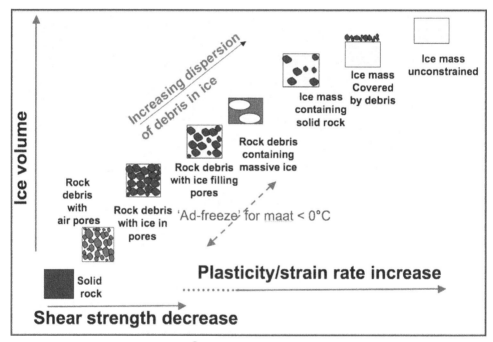

Fig. 4. This indicates both the possible mixture models likely to have been associated with the formation of most discrete debris associations and their mechanical properties. Thus, from the bottom left, rock slopes may fail and provide rigid blocks, perhaps amassed as scree with air spaces. If water/ice is mixed with the particles the mass is still rigid until there is enough ice in the mixture to allow ice deformation (Azizi and Whalley, 1995; Whalley and Azizi, 1994).

8. Interpreting Discrete Debris Accumulations

Table 1 (after Whalley, 2009) lists the main features likely to be seen as Younger Dryas Discrete Debris Accumulations in the uplands of the British Isles. This must, at present, be taken as a rather rough typology. It has not proved possible to provide a key system to help identify features. There are three reasons for this. First, the features themselves are somewhat variable in form and location on a hillside. Secondly, the debris input location and type needs to be taken into account (following from Figure 3). Thirdly, the interpretation itself may change. Thus, some of the following photographs show variations in form. The diverse papers about the origin of the Beinn Alligin 'rockslide' exemplifies both the second and third reasons starting with the original description (Sissons, 1975; Whalley, 1976) and with further detailed interpretations (Ballantyne, 1987; Ballantyne & Stone, 2004; Gordon, 1993). A clear example of a change in opinion is that by Wilson, already mentioned, in revising his formation model of some rock glaciers in northern Ireland to be massive rockslides. This view then casts doubt on the interpretation of the rock glacier (in the same geology) on Islay (Dawson, 1977). This also illustrates a further difficulty, that of terminology, a problem that has long bedevilled this area of research, especially that of rock glaciers (Hamilton & Whalley, 1995; Martin & Whalley, 1987).

Feature name	Comments on formation etc	Environmental interpretation use or caution
Blockfield *	If autochthonous (*in situ*): i Was it deformed by over-riding ice sheets? ii How old is it?	If undeformed or not removed how is this interpreted? Possible cosmogenic ratio exposure data. Use of tors related to blockfield might be helpful.
Hummocky moraine*	Passive formation (ablation) In some cases might be related to push moraines	Various interpretations, related to moraines, debris transport location, ice deformation; possible link to Østrem-type moraine
Landslide	Any YD or post-glacial event	Ice probably not involved but the resultant landform may look like one or other of the features listed here.
Østrem-type moraine	Originally, frontal debris deposition over 'old' snowbank; Possible confusion with: i Push moraine ii Rock glacier (glacier ice or permafrost) iii Protalus lobe iv Hummocky moraine	Relict feature difficult to interpret due to lack of ice and (as far as known) a significant relict feature. May look like a rock glacier - which then provides possible interpretation problems. To date, these have not been attributed to any feature in the British Isles.
Protalus lobe	i Involvement with glacier/snowbank ice + debris	Glacial, nival or permafrost maintenance, length of time of

Feature name	Comments on formation etc	Environmental interpretation use or caution
	input flux ii Involvement with permafrost-derived + ice debris input flux	preservation; dating problems possible
Protalus rampart	i Debris passively over snowbank ii Construction by small glacier iii Might develop into rock glacier (permafrost or glacier related?)	Size may indicate origin of ice; assumption that snow-derived relates to regional snowline rather than possible glacierisation altitude
Push moraine *	Topographic forms may have various origins and perhaps associated glacier dynamics	Interpretation as glacier margin movement or permafrost-related dynamics?
Rock glacier	i Glacier origin ii Permafrost origin iii Rockslide relict iv Composite origin v Breach of lateral moraine wall (not known in the British Isles)	Permafrost formative conditions or glacier; use in constructing regional trends for glacier ice (below regional limit) or assumption that all rock glaciers are of permafrost origin; Difficult to trace if rockfall-related
Rockslide	Large, one-off event, YD or post-glacial	Ice probably not involved but the resultant landform may look like one or other of the features listed here (see also 'landslide'). Bergsturz also used in the literature
Talus (scree slope)	Usually unambiguous; Length of time of formation may be considerable.	Paraglacial reactivation of old feature possible; may grade into other features down-slope (protalus lobe, protalus rampart, rock glacier)

Table 1. This table (derived from Whalley 2009), provides a summary of the main discrete debris accumulations likely to be found in the British Isles and Ireland, other than moraines. The features are listed alphabetically but those marked * are not referred to in this paper. Protalus lobe is equivalent to 'lobate rock glacier' or 'valley wall rock glacier' of some workers. A protalus ramparts is also known as 'winter nival ridge', 'pronival ridge' or 'snow-bed feature'.

9. Some examples of Discrete Debris Accumulations in the British Isles and possible analogues

The following illustrations illustrate some of the features in Table 1 as well as highlight interpretational problems associated with them. Where appropriate the UK National Grid co-ordinate system is used.

Fig. 6. Terminal area of a small glacier descending from Y Glydder, Snowdonia (SH 625727). This may have been a debris covered section of the lower glacier or even an incipient rock glacier.

Fig. 7. Corrie in Tröllaskagi, Northern Iceland where there has been a small glacier but very little debris to protect the ice from melting (Whalley, 2009). The debris cover is left as an indistinct trace after the ice has melted. A neighbouring corrie has a distinct rock glacier feature (Whalley et al., 1995) produced by high ice fluxes but with corresponding debris input to protect the ice.

Fig. 8. Protalus rampart, Herdus Scaw (NY 111161) one of several in the English Lake District described by Ballantyne and Kirkbride (1986). This is typical of those found in the uplands of the British Isles and is associated with snowpatch or snowbed. It is not known how long it took to build such a feature but a few hundred years seems a reasonable possibility.

Fig. 9. This protalus rampart (Oxford, 1985) is considerably larger than that shown in Figure 8 and has also been called a moraine (Sissons, 1980).

Fig. 10. The feature (arrowed) shown in Figure 9 in context of the north-facing cliffs of Robinson (NY 197176) English Lake District. Viewed like this it becomes easier to visualise a small glacier building the moraine/protalus rampart and why is it perhaps difficult to distinguish between the terms if they relate to the size of the snowbank/glacier. A similar example can be found below Fan Hir, Mynydd Du, South Wales (Shakesby & Matthews, 1993; Whalley & Azizi, 2003).

Fig. 11. Protalus rampart being formed by debris falling from the cliff and sliding/avalanching to the rampart feature. The rock is gabbro and would be equivalent to the low weathering rates from cliffs in Upland Britain during the Younger Dryas. Goverdalen, Lyngen Alps, Troms, Norway.

Fig. 12. Feature below Dead Crags, English Lake District (NY 267318) desribed as protalus rampart (Ballantyne & Kirkbride, 1986) and Oxford (1985). In contrast to the features shown in Figures 8 and 9 the deposit here is very subdued and tends more towards a lobe.

Fig. 13. Rock glacier or protalus lobe below the cliffs of Craig y Bera, Nantlle , Gwynedd, North Wales (SH 541538). This is an unusual feature in that it faces south although the cliffs here appear to weather more easily than other locations in the area. It may be a 'landslide' rather than rock glacier although may be similar to landslide features described by (Watson, 1962) some 20km south at Tal y Llyn.

Fig. 14. Feature (between arrow heads) interpreted as a (talus) rock glacier in the northern Lairig Ghru, Cairngorms, Scotland (NH961037) (Ballantyne et al., 2009; Sandeman & Ballantyne, 1996) and is similar to features found in Strath Nethy (see also Wilson 2009). The valley of the Lairig Ghru has many 'avalanche landforms' (Ballantyne & Harris 1994) which may be related, although these are much more down-slope, linear features.

Fig. 15. Although this has some similarities to protalus lobes this feature, on the edge of the Kinder Scout Millstone grit escarpment, Derbyshire (SK089895) is probably a slump/landslide. As it faces south-west snow/ice is unlikely to have lasted long here. However, ridges interpreted as moraines have been found at Seal Edge some 4km to the west where the escarpment is north facing (Johnson et al., 1990).

10. Conclusions

There are many features that may have been formed during the Younger Dryas (Loch Lomond Stadial) event in the British Isles. The examples shown here suggest that there are often different, and changing, opinions as to how they were formed and thus their environmental and climatic significance. The size of a ridge on a protalus rampart may be large enough to have been produced by a small glacier as opposed to a snowpatch. Mixing and matching the snow/ice/debris quantities and fluxes may produce a continuum of landforms and interpreting these forms presents problems. These mixtures may also have a significant effect on the mechanical properties, especially where ice is mixed with rock fragments. Landslides may well produce forms that look similar to forms such as protalus lobes and ridges. Although many of these forms have been mapped over thirty years or more, further work is needed to provide unequivocal interpretation of their formative mechanism and thus environmental significance. This is particularly important when the diversity of possible interpretations is viewed. Thus, slope failures (such as Fig. 15) may be indicative of the susceptibility of local geology to slope failure (which may be re-activated by localised human intervention) rather than of past climatic conditions. Conversely, in areas such as the British Isles where there is a strong west-east climatic gradient (Fig. 1), identification of features such as debris accumulations may assist in the evaluation of past climates and the extent of ice or perennial snow.

11. References

Andrews, J. T. (1975), *Glacial systems*, Duxbury, North Scituate, Ma.

Azizi, F. & W. B. Whalley (1995), Finite element analysis of the creep of debris containing thin ice bodies, Proc 5th International Offshore and Polar Engineering Conference, International Society of Offshore and Polar Engineers, The Hague. Vol.2, pp. 336-341.

Bakke, J.; S. O. Dahl, Ø. Paasche, R. Løvlie, & A. Nesje (2005), Glacier fluctuations, equilibrium line altitudes and palaeoclimate in Lyngen, Northen Norway, during the Lateglacial and Holocene, *The Holocene*, Vol.15, No.4, pp. 518-540.

Bakke, J.; Ø. Lie, S. O. Dahl, A. Nesje, & A. E. Bjune (2008), Strength and spatial patterns of the Holocene wintertime westerlies in the NE Atlantic region, *Global and Planetary Change*, Vol.60, No.1-2, pp. 28-41.

Ballantyne, C. K. (1987), The Beinn Alligin 'rock glacier', Quaternary Research Association, Cambridge, Wester Ross Field Guide.

Ballantyne, C. K. (2002), Paraglacial geomorphology, *Quaternary Science Reviews*, Vol. 21, No.18-19, pp.1935-2017.

Ballantyne, C. K. (2007), Paraglacial geomorphology, In: *Encyclopedia of Quaternary Science*, edited by S. A. Elias, pp. 2170-2182, Elsevier.

Ballantyne, C. K. & M. P. Kirkbride (1987), Rockfall activity in upland Britain during the Loch Lomond Stadial, *The Geographical Journal*, Vol.153, pp. 86-92.

Ballantyne, C. K. & C. Harris (1994), *The Periglaciation of Great Britain*, 330 pp., Cambridge University Press, Cambridge.

Ballantyne, C. K. & J. Stone (2004), The Beinn Alligin rock avalanche, NW Scotland: cosmogenic 10Be dating, interpretation and significance, *The Holocene*, Vol.14, No.3, pp. 448-453.

Ballantyne, C. K.; J. O. Stone, & L. K. Fifield (1998), Cosmogenic Cl-36 dating of postglacial landsliding at the Storr, Isle of Skye, Scotland, *The Holocene*, Vol.8, No.3, pp. 347.

Ballantyne, C. K.; C. Schnabel, & S. Xu (2009), Exposure dating and reinterpretation of coarse debris accumulations (rock glaciers) in the Cairngorm Mountains, Scotland, *Journal of Quaternary Science*, Vol.24, No.1, pp. 19-31.

Barsch, D. (1971), Rock glaciers and ice-cored moraines, *Geografiska Annaler*, Vol.53A, pp. 203-206.

Benn, D. I. & S. Lukas (2006), Younger Dryas glacial landsystems in North West Scotland: an assessment of modern analogues and palaeoclimatic implications, *Quaternary Science Reviews*, Vol.25, No.17-18, pp. 2390-2408.

Boardman, J. (Ed.) (1987), *Periglacial Processes and Landforms in Britain and Ireland*, pp. 296, Cambridge University Press, Cambridge.

Chorley, R. J.; A. J. Dunn, & R. P. Beckinsale (1973), *The History of the Study of Landforms: or the Development of Geomorphology: Volume 1 : Geomorphology before Davis*, pp. 678, Methuen, London.

Chorlton, J. C. & H. Lister (1971), Geographical controls of glacier budget gradients in Norway, *Norsk Geografisk Tidsskrift*, Vol.25, pp. 159-164.

Church, M. & J. M. Ryder (1972), Paraglacial sedimentation: a consideration of fluvial processes conditioned by glaciation, *Bulletin of the Geological Society of America*, Vol.83, pp. 3059-3072.

Clark, C. D.; P. L. Gibbard, & J. Rose (2004), Pleistocene glacial limits in England, Scotland and Wales, *Developments in Quaternary Science*, Vol.2, pp. 47-82.

Davies, M. C. R.; O. Hamza, & C. Harris (2003), Physical modelling of permafrost warming in rock slopes, Proceedings of the 8th International Permafrost Conference, Swetts and Zeitlinger, Zürich.

Dawson, A. G. (1977), A fossil lobate rock glacier in Jura, *Scottish Journal of Geology*, Vol. 13, pp. 31-42.

Farr, T. G. (2004), Terrestrial analogs to Mars: The NRC community decadal report, *Planetary and Space Science*, Vol.52, No.1-3, pp. 3-10.

Gillen, C. (2003), *Geology and landscapes of Scotland*, Terra, Harpenden, Hertfordshire.

Golledge, N. R. (2010), Glaciation of Scotland during the Younger Dryas stadial: a review, *Journal of Quaternary Science*, Vol.25, No.4, pp. 550-566.

Gollege, N. R. & A. Hubbard (2005), Evaluating Younger Dryas glacier reconstructions in part of the western Scottish Highlands: a combined empirical and theoretical approach, *Boreas*, Vol.34, pp. 274-286.

Gordon, J. E. (1993), Beinn Alligin, in *Quaternary of Scotland*, Edited by J. E. Gordon and D. G. Sutherland, pp. 118-122, Joint Nature Conservation Committee, Chapman and Hall, London.

Gordon, J. E. (2006), Shaping the landscape, In *Hostile Habitats; Scotland's Mountain Environment*, Edited by N. Kempe and M. Wrightham, pp. 52-83.

Gordon, J. E. & D. G. Sutherland (1993), *Quaternary of Scotland*, Chapman and Hall, London.

Hamilton, S. J. & W. B. Whalley (1995), Rock glacier nomenclature: a re-assessment, *Geomorphology*, Vol.*14*, pp. 73-80.

Harrison, S. & E. Anderson (2001), A Late Devensian rock glacier in the Nantlle valley, North Wales, *Glacial Geology and Geomorphology*.

Harrison, S.; E. Anderson, & D. G. Passmore (1998), A small glacial cirque basin on Exmoor, Somerset, *Proceedings of the Geologists' Association*, Vol.109, pp. 149-158.

Harrison, S.; B. Whalley, & E. Anderson (2008), Relict rock glaciers and protalus lobes in the British Isles: implications for Late Pleistocene mountain geomorphology and palaeoclimate, *Journal of Quaternary Science*, 23(3), 287-304.

Hauber, E.; Reiss, D., Ulriuch, M., Preusker, F., Trauthan, F., Zanetti, M., Hiesinger, H., Jaumann, R., Johansson, L., Johnsson, A., Van Gasselt, S., & Olvmo, M.. (2011), Landscape evolution in Martian mid-latitude regions: insights from analogous periglacial landforms in Svalbard, in *Martian Geomorphology*, Edited by M. R. Balme, A. S. Bargery, C. J. Gallagher & S. Gupta, pp. 111-131, Geological Society, London.

Hubbard, A. (1999), High-resolution modelling of the advance of the Younger Dryas ice sheet and its climate in Scotland, *Quaternary Research*, Vol.*52*, pp. 27-43.

Ivy Ochs, S.; C. Schlüchter, P. W. Kubik, & G. H. Denton (1999), Moraine exposure dates imply synchronous Younger Dryas glacier advances in the European Alps and in the Southern Alps of New Zealand, *Geografiska Annaler: Series A, Physical Geography*, Vol.81, No.2, pp. 313-323.

Jarman, D. (2006), Large rock slope failures in the Highlands of Scotland: characterisation, causes and spatial distribution, *Engineering Geology*, Vol.83, No.1-3, pp. 161-182.

Jarman, D. (2009), Paraglacial rock slope failure as an agent of glacial trough widening, *Periglacial and Paraglacial Processes and Environments*, edited by J. Knight and S. Harrison, Geological Society, London. Special Publication 320, pp. 103-131.

Johnson, R. H.; J. H. Tallis, & P. Wilson (1990), The Seal Edge Coombes, North Derbyshire: a study of their erosional and depositional history, *Journal of Quaternary Science*, Vol.5, No.1, pp. 83-94.

Martin, H. E. & W. B. Whalley (1987), Rock glaciers. Part 1: rock glacier morphology: classification and distribution, *Progress in Physical Geography*, Vol.11, No.2, pp. 260-282.

McKirdy, A.; J. E. Gordon, & R. Crofts (2007), *Land of Mountain and Flood*, Birlinn/Scottish Natural Heritage, Edinburgh.

Meissl, G. (1998), Modelierung der Reichweite von Felsstürzen. Fallbeispiel zur GIS-gestützen Gefahrenburteillung aus dem Bayrisches und Tiroler Alpenraum. *Rep.*, pp. 249, Innsbruck.

Menounos, B. & M. A. Reasoner (1997), Evidence for Cirque Glaciation in the Colorado Front Range during the Younger Dryas Chronozone, *Quaternary Research*, Vol.48, No.1, pp. 38-47.

Muscheler, R.; B. Kromer, S. Bjorck, A. Svensson, M. Friedrich, K. Kaiser, and J. Southon (2008), Tree rings and ice cores reveal 14C calibration uncertainties during the Younger Dryas, *Nature Geoscience*, Vol.1. No.4, p.263.

Nakawo, M.; Raymond, C.F. & Fountain, A. (Eds.) (2000), *Debris covered glaciers*, 288+viii pp., IASH publ 264.

Oldroyd, D. R. (2002), *Earth, Water, Ice and Fire: two hundred years of geological research in the English Lake District*, The Geological Society, London.

Østrem, G. (1964), Ice-cored moraines in Scandinavia, *Geografiska Annaler*, 46A, 282-237.

Østrem, G. (1971), Rock glaciers and ice-cored moraines, a reply to D. Barsch, *Geografiska Annaler*, 53A, 207-213.

Oxford, S. P. (1985), Protalus ramparts, protalus rock glaciers and soliflucted till in the northwest part of the English Lake District, in *Field Guide to the Periglacial Landforms in the English Lake District*, edited by J. Boardman, Quaternary Research Association, Cambridge, 38-46 pp.

Rose, J. (1985), The Dimlington Stadial/Dimlington Chronozone: a proposal for naming the main glacial episode of the Late Devensian in Britain, *Boreas*, 14(3), 225-230.

Rudwick, M. J. S. (2008), *Worlds before Adam*, University of Chicago Press, London.

Sandeman, A. F., and C. K. Ballantyne (1996), Talus rock glaciers in Scotland: characteristics and controls on formation, *Scottish Geographical Journal*, 112(3), 138-146.

Shakesby, R. A. (1997), Pronival (protalus) ramparts: a review of forms, processes, diagnostic criteria and palaeoenvironmental implications, *Progress in Physical Geography*, 21, 394-418.

Shakesby, R. A., and J. A. Matthews (1993), Loch Lomond Stadial glacier at Fan Hir, Mynydd Du (Brecon Beacons), South Wales: critical evidence and palaeoclimatic implications, *Geological Journal*, 28, 69-779.

Shakesby, R. A., A. G. Dawson, and J. A. Matthews (1987), Rock glaciers, protalus ramparts and related phenomena, Rondane, Norway: a continuum of large-scale talus-derived landforms, *Boreas*, 16, 305-317.

Sissons, J. B. (1975), A fossil rock glacier in Wester Ross, *Scottish Journal of Geology*, 92, 182-190.

Sissons, J. B. (1980), The Loch Lomond Advance in the Lake District, northern England, *Transactions of the Royal Society of Edinburgh: Earth Sciences*, 71, 13-27.

Slaymaker, O. (2007), Criteria to discriminate between proglacial and paraglacial environments, *Landform Analysis*, 5, 72-74.

Slaymaker, O. (2009), Proglacial, periglacial or paraglacial?, in *Periglacial and Paraglacial Processes and Environments*, edited by J. Knight and S. Harrison, Geological Society, London. Special Publication 320, pp. 71-84.

Watson, E. (1962), The glacial morphology of the Tal-y-Llyn Valley, Merionethshire, *Transactions and Papers Institute of British Geographers*(30), 15-31.

Whalley, W. B. (1976), A fossil rock glacier in Wester Ross, *Scottish Journal of Geology*, 12, 175-179.

Whalley, W. B. (1984), Rockfalls, in *Slope instability*, edited by D. Brunsden and D. B. Prior, pp. 217-256, Wiley, Chichester.

Whalley, W. B. (2004), Glacier research in mainland Scandinavia, in *Earth paleoenvironments: records preserved in mid- and low-latitude glaciers*, edited by L. D. Cecil, J. R. Green and L. G. Thompson, pp. 121-143, Kluwer, Dordrecht.

Whalley, W. B. (2009), On the interpretation of discrete debris accumulations associated with glaciers with special reference to the British Isles in Periglacial and Paraglacial Processes and Environments, edited by J. Knight and S. Harrison, *Geological Society, London, Special Publications, 320*, pp. 85-102.

Whalley, W. B. & Martin, H. E. (1992), Rock glaciers: II model and mechanisms, *Progress in Physical Geography, 16*, 127-186.

Whalley, W. B., and Azizi, F. (1994), Models of flow of rock glaciers: analysis, critique and a possible test, *Permafrost and Periglacial Processes*, Vol.5, pp. 37-51.

Whalley, W. B., and Azizi, F. (2003), Rock glaciers and protalus landforms: analogous forms and ice sources on Earth and Mars, *Journal of Geophysical Research, Planets*, Vol. 108(E4), 8032, (DOI: 8010.1029/2002JE001864).

Whalley, W. B.; Douglas, G. R. & Jonsson, A. (1983), The magnitude and frequency of large rockslides in Iceland in the Postglacial, *Geografiska Annaler, 65A*, 99-110.

Whalley, W. B.; Palmer, C. F., Hamilton, S.J. & Kitchen, D. (1996), Supraglacial debris transport, variability over time: examples from Switzerland and Iceland, *Annals of Glaciology*, Vol.22, pp. 181-186.

Whalley, W. B., Hamilton, S. J., Palmer, C. F., Gordon, J. E., & Martin, H. E. (1995), The dynamics of rock glaciers: data from Tröllaskagi, north Iceland, in *Steepland Geomorphology*, Edited by O. Slaymaker, pp. 129-145, Wiley, Chichester.

Wilson, P. (2004), Relict rock glaciers, slope failure deposits, or polygenetic features? A reassessment of some Donegal debris landforms, *Irish Geography*, Vol.37, pp. 77-87.

Wilson, P. (2009), Rockfall talus slopes and associated talus-foot features in the glaciated uplands of Great Britain and Ireland in Periglacial and Paraglacial Processes and Environments, edited by J. Knight and S. Harrison, *Geological Society, London, Special Publications*, Vol.320, pp. 133-144.

Wilson, P. (2010), *Lake District Mountain Landforms*, Scotforth Books, Lancaster.

4

Environmental Changes in Lakes Catchments as a Trigger for Rapid Eutrophication – A Prespa Lake Case Study

Svetislav S. Krstić

Faculty of Natural Sciences, St.Cyril and Methodius University, Skopje,
Macedonia

1. Introduction

Elucidating upon and/or separating the natural processes of eutrophication from the anthropogenically induced ones in a lake's history have proven to be a formidable task. The nature and patterns of the eutrophication processes, their overall impact on the ecosystem and biota, as well as the possible management practices to be introduced to reverse or slow down the accelerated eutrophication have been in focus only very recently mainly due to imposed EU legislation such as the WFD. Another important and also very demanding task that greatly influences management plans, costs and activities is the detection of the reference conditions for every particular water body. Among various suggested approaches, we have concluded that the best way is to reveal the past changes of the lake's environment by conducting paleo-ecological research (the so called *state change approach*) using core sample analyses of as many parameters as possible and relate them to the biota, algae in particular. In that regard, under the comprehensive River Basin Management Plan developed for the Prespa Lake catchment (the part that belongs to Macedonia), the most wide-ranging 12 month surveillance monitoring has been conducted in order to reveal the present ecological situation and the past changes during the last 10 ka period. The results of these investigations are presented in this chapter.

2. Investigated area

The Prespa Lake has been chosen as a part of the complex Prespa-Ohrid-River Crni Drim system which is thought to be more than 3-5 million years old, and as a pilot project as its catchment is relatively small. Nevertheless, obtaining the research objectives was far from simple, since the Prespa region includes two inter-linked lakes, namely the Micro Prespa and Macro Prespa, surrounded by mountainous ecosystems. These two lakes together form the deep points of an inner-mountainous basin that has no natural surface outflow. Drainage is only provided by means of underground links by which water of the Macro Prespa Lake (approximately 845 m asl) drains towards the west to the approximately 150 metres lower Ohrid Lake. On its northern shore, in the town of Struga, the Ohrid Lake has a natural outlet into the Crni Drim River. The Micro Prespa Lake is shared between Greece and Albania, while the Macro Prespa Lake is shared between Albania, the Republic of

Macedonia and Greece. The Ohrid Lake again belongs partly to the Republic of Macedonia and partly to Albania. The Micro and Macro Prespa Lakes are connected by a small natural channel here referred to as the Isthmus of Koula. Since 1969/70 the water level of the Micro Prespa has been controlled by a regulating weir structure to limit the maximum water level.

Fig. 1. Google map of the Prespa and Ohrid Lakes system, and the position of the investigated area in the map of Europe.

2.1 Geological and geomorphological features

According to Micevski (2000), the Ohrid-Prespa region is characterised by fairly complex geological-tectonic structures with rocks from the oldest Paleozoic formation to the youngest Neogene and Quaternary sedimentary rocks.

Prespa Valley is surrounded by the mountains of Petrinska Planina, Galichica, Suva Planina, Ivan Planina and Suva Gora. Both mountains and the valley are mainly composed of rocks varying in their age, mineralogical composition and origin. The calcareous rocks are dominant and to a small extent are distributed between magmatic rocks and Grano-Diorites. Syenites are present at the higher elevation areas, but Triassic carbonate rock masses are also present in many areas. Different types of Quaternary sediments, such as alluvial, fluvio-glacial, proluvial, organogenic-marsh and diluvial sediments, are dominant in the valley, especially at the riverbeds. The depth of those sediments varies between 100 and 200 m.

Since the carbonate rocks dominate the geology of Prespa Valley and it is of principal importance for the arguments presented in this chapter, the geomorphological features of Galichica Mountain (located on the North-West side of the valley – Fig. 2) will be presented in more details.

According to its morphometrical characteristics (hypsometry, exposition, inclination), Galichica Mountain exhibits a profound tectonical character (a horst) elevated between two lake basins (Fig. 3). Resulting from its geological composition (almost totally presented by Triassic carbonate rocks), hypsometrical features (only 7.19 km² of its surface is above 2.000 m elevation) and climate (or more its fluctuations in the past), the dominant morphogenetic processes that have caused and created modern relief on this mountain are karstic, glacial and periglacial. These processes have been intermingled (supplemented), changed in intensity and duration or fully stopped over the time due to climatic changes.

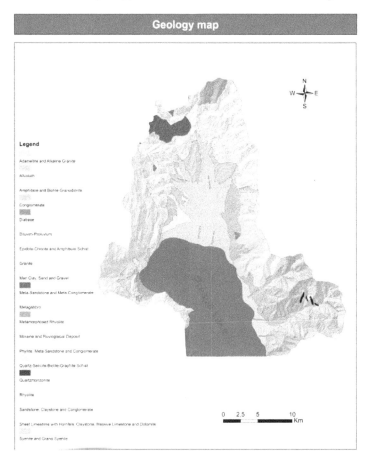

Fig. 2. Geological map of Prespa Lake (Macedonian part only).

Fig. 3. Galichica Mountain – a horst between two lakes, Prespa Lake on the left and Ohrid Lake on the right.

The *karstic features* are the dominant genetic type of relief forms on Galichica Mountain, which is a typical karstic area where the Triassic massive and banked limestone layers lie over the crystal shales. These surfaces have been exposed for a long time to the influence of external factors which have strongly initiated the process of karstification. Micro and macro relief karstic forms, such as *karrens*, numerous karst *sinkholes* and karstic *dry flows*, as well as *karstic fields*, are frequent. From the underground karstic forms, a dozen caves and two chasms have been registered.

Fig. 4. The peak Magaro (2.254 m), the karstic field Lipova Livada and the glacial cirque between them.

Mountain Galichica's altitude and its favourable morpho-plasticity enabled the accumulation of snow and ice during Pleistocene resulting in glacial relief formation. Aside from the two cirques (Fig. 4), a small number of cribs can also be found, but the dominant landscape is formed by the *periglacial* processes resulting in *stony horseshoes, slided blocks, grassy terraces, loose glacial residues* etc. (Fig. 5).

In relation to water transport and balance, such a complex geomorphology based on karstic fundaments has created the only possible type of aquifer in the mountain region – the *fractured type of aquifer* with medium or low water permeability (yield between 0.1 and 5.0 L·s^{-1}). The most presented rocks in this aquifer are Paleozoic schists with a low degree of crystallisation (chlorite, sericite, quartz schist and quartzite) Less presented are gneiss, mica-schist and amphibole schist, as well as intrusive rocks (gabbro, granite, syenite, diorite, diabase, quartz-porfire, andesite, etc.).

All those rock masses are rugged with faults, fractures and cracks, which generally cannot accumulate larger reserves of groundwater. These rocks have a very low coefficient of filtration (kf < E-4 cm·s^{-1}) and different intensity of fissuring. Local fissuring is very important due to the presence of intense subsurface fissuring and surface to sub-surface weathering (locally over 20 m depth), which influence the hydro-geological characteristics of the rock masses. Often the above mentioned fractures and cracks are closed and filled up with clay material.

Fig. 5. Periglacial landscape on Galichica Mountain.

Similarly the granites, granodiorites and syenites are characterised with low water permeability, with yield of 0.1 – 0.5 L cm·s⁻¹ (Pelister, Ilinska mountain, Stogovo, etc.).

Having this kind of natural geomorphology in a lake's catchment means that any significant change in the land cover would result in massive washout of nutrients, minerals and sediments into the lake, due to:

- dominant surface run-off and negligible penetration of the water into the underground aquifers;
- increased erosion (*erosion risk* – Fig. 6) due to abrasive forces on exposed carbonate rocks;
- decreased capacity of soils to accumulate and store excess amounts of nutrients.

2.2 Hydrology

The Prespa catchment area includes two lakes: Micro and Macro Prespa, and permanent or seasonal streams, which discharge into the two lakes. The major contributing rivers to Macro Prespa Lake are Golema, Brajcinska and Kranska Rivers in the Republic of Macedonia and Aghios Germanos River in Greece. There is no major source of surface water input from Albania to Mikro Prespa. Together with the Ohrid Lake, the Prespa Lakes form a unique ensemble of water bodies in the Balkan regions. The Prespa Lakes form the deep points of an inner-mountainous basin that has no natural surface outflow. Drainage is only provided through karstic underground links by which water of the Macro Prespa Lake (approximately 845 m asl) drains westwards towards the Ohrid Lake lying approximately

150 metres below. The Macro Prespa and the Ohrid lakes are separated by the Galichica Mountain, where a surface drainage network is almost absent due to the karstic nature of that mountain chain. The karstic mountains of Mali i Thate (Dry Mountains) in Albania and Galicica in are highly porous, resulting in high capacity for water transport.

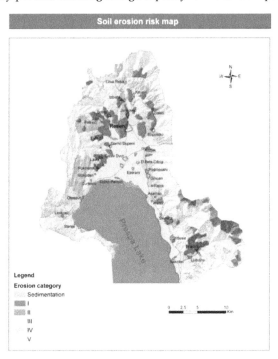

Fig. 6. Soil erosion risk map for the Prespa Lake watershed (Macedonian part only).

The karstic connection of the Prespa Lake and Ohrid Lake is evidenced by spring observations. The Saint Naum spring and Tushemisht springs on the south coast of the Ohrid Lake originate from the Prespa Lake. Lake Ohrid has a surface outflow to the Crni Drim River. Since 1962 the outflows of the Ohrid Lake have been controlled in the City of Struga by means of a weir. Operation of the weir is governed by the needs of downstream hydropower plants and the need to maintain the lake's water level within optimal levels (maximal water level is 693 m and the minimum water level is 691.65 m).

The Micro and Macro Prespa Lakes are connected in Greece via a small natural channel. From the North and East of the catchment, several small and mostly ephemeral watercourses flow into the Macro Prespa Lake.

The main tributary of Prespa Lake is the Golema River. Two springs which are situated on the slopes of Plakenska and Bigla Mountains form the water courses of the Leva and Bukovska Rivers that join into the Golema River. Elevations of the Golema River springs are at 1440.00 m asl, while it inflows into the lake at 850.00 m asl. Larger tributaries of Golema River flow on the left side of the catchment. These are Krivenska, Celinska and Sopotska. Bolnska River used to inflow into the Golema River in the past, while at present it has become a tributary of Istocka River which inflows into the lake. The total catchment area of the Golema River is F=182.9 km², the length of the river is L=26.1 km.

During the last century the Macro Prespa Lake experienced a significant water level fluctuation. After its last peak in June 1963, the level generally dropped to its last low by approximately 8 metres up until 2002 (Fig. 7). This signifies a loss in water volume, with a loss of water surface maybe being even more significant. The major tributary of Macro Prespa Lake, the Golema River, shows high variations of the discharge during that time. It has been noticed that for the series of the minimal discharges it is typical that in some years, during the summer period, the river has very low water discharges or even no water at all. Floods have been registered in 1962 and 1979 with maximal discharge of Q max=33.4 m³/s and 30.0 m³/s respectively.

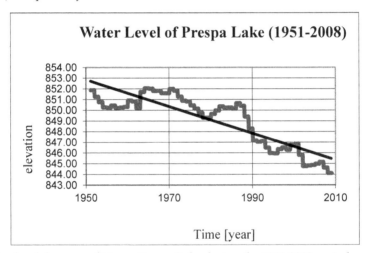

Fig. 7. Water level decrease of Macro Prespa Lake during the 1951-2008 period.

2.3 Soils

Depending on the various pedogenetic factors in the region, there are several soil types presented on the soil map (Fig. 8). Dominant soils in the Prespa Valley are *alluvial soils* (fluvisols) located in the lowest region. These soils are formed over the sediments by the rivers. On the other hand, on the central region of the Ezerani protected area, as well as within the band closest to the lakeshore – *hydric soil* formation is ongoing, which leads to the formation of *gleysols* in different stages of evolution. Around the fluvisols, *colluvial soils* are well developed. These soils are formed above thicker sediments and are being created by the rivers and torrents in the area. On a significant part in the valley and hills on the western side, *chromic luvisols* have been formed and these soils are mainly used for agriculture. On the mountain region various types of *cambisols* have been formed. On the Baba Mt., *eutric* and *district cambisols* are dominant. Natural vegetation adapts to the soils and on the eutric cambisols there are oak stands, whilst on the district cambisols beech compositions stand. *Rankers* (humus-accumulative soils) with various phases of development are formed on the highest altitude in the subalpine and alpine areas. On these soils only grass vegetation is growing.

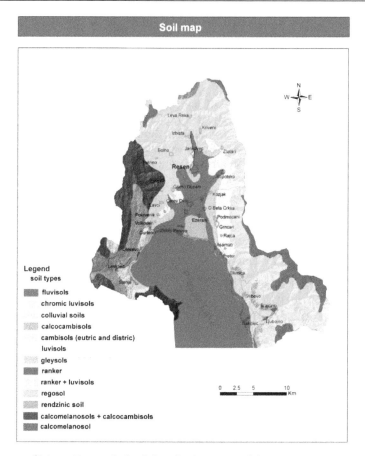

Fig. 8. Soil map of Macro Prespa Lake (Macedonian part only).

2.4 Climate

Being close to the Mediterranean seas (Adriatic and Aegean), it could be expected that the influence of the Mediterranean climate on the location of Prespa Valley would be significant. But the high mountains surrounding the valley mark the existing highland properties of the climate. The specific orographic conditions that have an impact on the dynamic factors of the climate, as well as the impact of geographical and local factors, create three different types of climate throughout the whole watershed (Fig. 9). The whole region of the Prespa Lake watershed belongs to the Eco-region 6 – Hellenic Western Balkan.

- Warm and cold sub-Mediterranean climatic area, from 600 to 900 m and from 900 to 1,100 m altitude, respectively,
- Sub-mountainous and mountainous sub-Mediterranean climatic area, from 1,100 m to 1,300 m and from 1,300 m to 1,650 m altitude, respectively,
- Sub-alpine and alpine climatic area from 1,650 m to 2,250 m and above 2,250 m altitude, respectively.

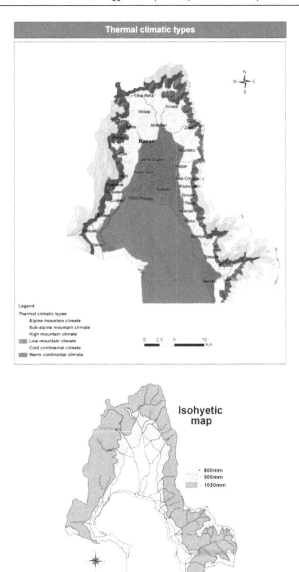

Fig. 9. Climatic types and isohyetic map of Macro Prespa Lake (Macedonian part only).

2.5 Flora and fauna

The vegetation varies from submerged aquatic formations and reed beds to shrublands of junipers and oaks to forests of oak, beech, mixed broadleaves to alpine grassland. In total, there are 1,326 plant species in Prespa, 23 freshwater fish species, 11 amphibian and 21 reptile species, more than 42 mammal species, among which are the brown bear, the wolf, the otter and the chamois, and over 260 bird species.

A shelter for over 90 species of migratory birds, Prespa Lakes are also home to tens of species that have been officially registered as critically endangered or vulnerable. Among them is the Dalmatian Pelican, one of the largest flying birds in the world, who seeks secluded wetlands to build nests and to hatch chicks in what is its largest breeding colony worldwide.

Fig. 10. Dominant reed bed flora and the Dalmatian Pelican (*Pelicanus crispus*) on Macro Prespa Lake.

From the phytocoenological point of view, the presence of endemic plant community *Lemneto-Spirodeletum polyrrhize aldrovandetosum* is the most important.

3 Anthropogenic impacts on the status of surface and ground water bodies

3.1 Estimation of the point source of pollution

In the Municipality of Resen, according to the 2002 census of the Republic of Macedonia, 16,825 people live in 44 inhabited places. In addition there are several tourist centres that are creating additional pressure on the sewage network and water bodies, especially in the summer period: Hotel Pretor, Pretor – 54 guests and 50 summer houses (around 200 persons = 254 people/total); Hotel Kitka, Resen, 40 guests in total; Auto camp Krani, house trailers and tents 3,000 beds, 42 villas 168 beds and 32 bungalows 130 beds = 3,298 people/total; private accommodation in villages Brajčino, D.Dupeni, Pretor, Slivnica, Ljubojno and Stenje = 375 tourists. The calculations for this magnitude of pressure are presented on Tab. 1.

According to these calculations, current load from household sewage (without wastewater treatment) plays a significant role in the pollution of water bodies in Prespa Lake watershed.

On the Macedonian side of Prespa Lake there are several medium-size enterprises from eight industrial branches: food processing, poultry farming, textile, metal processing, wood processing, civil construction, ceramics and the chemical industry.

Based on the environmental data provided by the enterprises themselves, an overall estimation of the point source pollution can be determined as presented in Tab. 2 (values of considerable pollution are coloured in red):

Parameter	Unit	Value
Inhabitants	person	20,792
Q~water~ per capita	l/d*People Equivalent	150
BOD$_5$	g/PE*d	60
COD	g/PE*d	110
TSS	g/PE*d	70
N (as TKN)	g/PE*d	8.8
P	g/PE*d	1.8
Calculation for Wastewater Quantity and Quality: [1]		
Flow (Q)=(People*Q per capita)/1000	m³/d	3,118.8
	m³/year	1,138,362
BOD$_5$	kg/d	1,247./5
	kg/year	455,344.8
	mg/l	400
COD	kg/d	2,287.1
	kg/year	834,798.8
	mg/l	733.3
TSS	kg/d	1,455.4
	kg/year	531,235.6
	mg/l	466.7
N	kg/d	183
	kg/year	66,783.9
	mg/l	58.7
P	kg/d	37.4
	kg/year	13,660.3
	mg/l	12

Table 1. Calculations for 20,792 people (together with 3,967 tourists), based on average loads per person.

---

[1] Based on German ATV 131A Standard (May, 2000) and German Wastewater Ordinance (2004) for assessing wastewater load, approved by Guidance for the Analysis of Pressures and Impacts in accordance with the WFD (2004).

Indicator:	SwissLion (Agroplod) doo (5.11.2008) 3rd point (biscuits-napolitana	SwissLion (Agroplod) doo (5.11.2008) 2nd point (resana cakes)	SwissLion (Agroplod) doo (5.11.2008) 1st point (coffee & peanuts)	Algreta AD Resen (14.10.2009) Recipient Golema River	CD Frut, Carev Dvor (28.11.2008) Recipient Bolsnica River	MDK (II class waters)*	Total:
pH value	6.5	6.5	8.7	6.54	6.2	6.5- 6.3	6.88
Total suspended solids TSS (mg/L)	25	30	25	29	53	10 – 30	162
BOD₅ (mg/L)	4.5	6.6	7.3	7.7	5.3	2 – 4	31.4
COD (mg/L)	341	372	341	18.4	9	2.5 – 5	1.081
Nitrates (mg/L)	3	50	3	0.4	1.3	15	57.7
Nitirites (mg/L)	0	0	0	0	0.3	0.5	0.3
NH₄ (mg/L)	0.4	0.150	0	0.19	0.1	0.02	0.84
Fe (mg/L)	/	/	/	>1	0.25	0.3	1.25
Mn (mg/L)	/	/	/	0.315	0.3	0.05	0.615
Al (mg/L)	/	/	/	0.009	/	1-1.5	0.009
Cd (mg/L)	/	/	/	/	0.0005	0.0001	0.0005
Cl₂ (mg/L)	14.9	17.7	82.2	/	0.0025	0.002	114.8
Cr total (mg/L)	/	/	/	/	0.038	0.05	0.038
Cu (mg/L)	/	/	/	/	0.012	0.01	0.012
Ni (mg/L)	/	/	/	/	0.035	0.05	0.035
Zn (mg/L)	/	/	/	/	0.075	0.1	0.075
Turbidity (NTU)	20	10	20	393	/	0.5-1	443
Total N (mg/L)	/	/	/	/	/	0.2 -0.32	
TDS (mg/L)	385	290	580	/	146	500	1,401
Total P (mg/L)	/	/	/	/	/	10 – 25	
Eutrophication Indicators – Most probable number of thermo-tolerant coliform bacteria /100 ml	240,000	240,000	240,000	/	/	5 – 50	240,000

Table 2. Calculation of various pollutants per source of pollution.

From a poultry farm, typical emissions to wastewater include: ammonia, uric acid, magnesium, sulphates, total nitrogen (N) and total phosphorus (P), as well as small concentrations of heavy metals (Cu, Cr, Fe, Mn, Ni, Zn, Cd, Hg and Pb).

*According to the Regulation for Classification of Water (Official Gazette of the Republic of Macedonia No. 18-99)

Using these emission factors, total releases of NH_3 from manure in "Swiss Lion Agrar" poultry farm are: 13,600 kg/year. Some 720 mg·L^{-1} of the total nitrogen and total phosphorus concentrations of 100 mg·L^{-1} are released on average per year. The BOD levels are reported to be 1,000 – 5,000 mg·L^{-1} (BREF, 2003).

Process wastewater is a major source of pollutants from textile industries (WHO, 1993). It is typically alkaline and has high BOD, from 700 to 2,000 milligrams per litre, and high chemical oxygen demand (COD), at approximately 2 to 5 times the BOD level. The wastewater also contains chromium, solids, oil and possibly toxic organics, including phenols from dyeing and finishing, and halogenated organics from processes such as bleaching. Dye wastewaters are frequently highly coloured and may contain heavy metals such as copper and chromium. Wool processing may release bacteria and other pathogens as well. Pesticides are sometimes used for the preservation of natural fibres and these are transferred to wastewaters during washing and scouring operations. Pesticides are used for mothproofing, brominated flame retardants are used for synthetic fabrics and isocyanates are used for lamination.

Point sources of pollution in the watercourses arrive from domestic sewage networks, WWTP, sparsely built-up areas, industry, contaminated sites or storm water outfalls. In addition, there are other point source pressures like fish farming.

1. *Domestic wastewater (household sewage) load estimations based on pressure from 20,792 people (without WWT):* BOD$_5$ – 455,344.8 kg/year; COD - 834,798 kg/year; total suspended solids – 531,235.6 kg/year; N - 66,783.9 kg/year; P – 13,660.3 kg/year. Only 55% of settlements are connected to the domestic wastewater system (data only for Golema River basin). In the whole Prespa watershed percentage is even lower, due to the high amount of sparsely built-up areas.

2. *Industrial pollution*
   On the Macedonian side of Prespa Lake there are several medium-size enterprises from eight (8) industrial branches: food processing, poultry farming, textile, metal processing, wood processing, civil construction, ceramics and the chemical industry, all causing pressure on the water bodies.

   Major pressures are coming from industry facilities "Algreta" (aluminium & zinc foundry), "Hamzali" (ceramic plant), food and beverage industry "CD Fruit", "Swiss Lion Agroplod" (food and beverages) and "Swiss Lion Agrar" (poultry farm). Impacts on the water body include high levels of ammonium, nitrates, phosphorus, aluminium, very high concentrations of Cl_2, high BOD$_5$ and COD concentrations, increased number of thermo-tolerant coliform bacteria, increase in heavy metals pollution-Fe, Zn, Cr, Cd, very high turbidity, phenols, benzene, halogenated organics, illegal pesticides in high quantities, brominated flame retardants, isocyanates used for lamination, oils and grease. Both "Swiss Lion Agroplod" and "CD Fruit" Carev Dvor are planning to have their small WWTP operational in the near future, but currently they are discharging effluents directly into the water bodies (no pre-treatment). Wastewater coming from industrial facilities only in the town of Resen is estimated to be approximately 69,350 m^3/year. Total annual amount of wastewater from CD Fruit is around 9,000 m^3. There is also pressure from agricultural activities and from sparsely built-up areas and storm water outflows that do not have their own infrastructure.

3.2 Estimation of diffuse sources of pollution

Based on information given by representatives of the Union of Agricultural Associations and the local AES office, by and large fertilisation of apples/fruits among individual growers in the Prespa region is performed in three phases, as follows (Tab. 3):

- autumn basic fertilisation with complex NPK (4:7:28) fertiliser in amounts of 500 to 700 kg per hectare (kg·ha-1);
- early spring fertilising with complex NPK (15:15:15) in amounts of 400 to 600 kg·ha-1; and
- late spring fertilisation with usage of nitrate fertilisers, such as ammonium nitrate, in amounts of 300 to 400 kg·ha-1.

Some farmers limit fertiliser application to only twice per year. Use of organic fertilisers is very rare. Based on these data, the total annual quantity of fertilisers used for apple production in the Golema River basin (for 1,200 ha) equals roughly 1,900 tons. There is no information on fertiliser use for other crop types, but this type of data is not of interest because farmers are not growing anything other than apples and the future trend is in increased apple production. Nevertheless, the presented fertilisation scheme should be regarded as a mere generalisation used for approximation purposes and presented as the total calculated quantity.

In total 920,150 kg of nitrogen as nitrogen fertilisers is applied each season. It is practically impossible to determine to what extent farmers in the region overuse fertilisers. Furthermore, some of the more important general characteristics for the soil types found in the region are that mechanical content of all types with high percentage is sandy and with dominance of grit fractions, which means that the soil types are permeable for water and dissolved mineral matters. Therefore, water from precipitation and/or irrigation can exert a strong impact on the dilution of nitrogen forms from the fertilisers and other materials, which can finally reach the water courses by underground leaching or surface run-off.

System of Fertilisation and Period	Fertiliser Type	Quantity (kg/ha)	Active substances (kg/ha)		
			N	P_2O_5	K_2O
Basic autumn fertilisation	NPK 4:7:28	700	28	49	196
Early spring fertilisation	NPK 15:15:15	500	75	75	75
Late spring fertilisation	NH_4NO_3 34%	400	136	0.0	0.0
Total		1600	239	124	271

Table 3. Practice of fertilisation in private orchards in the Prespa region.

In total about 477 tons of phosphorous in P_2O_5 form is used in the catchment area. As a result of the widely accepted perception of low fertility of the soil with available phosphorous, a lot of P-fertilisers are used. Examples have been reported that farmers who have analysed soil samples in various soil-testing laboratories in the country have been advised not to apply certain nutrients – in particular P and K – for a period of three to four years in order to reach the required balance. Yet again, this cannot be taken as a general rule for the entire region, since there are farmers who, due to limited finances, do not use high quantities of fertilisers. Nevertheless, there is significant proof of overuse of phosphorous and it should be assumed as one of the major risks for pollution and eutrophication of the water from agricultural sources. More than 1,000 tons of potassium oxide is also applied through the fertilisation process in the Prespa region.

There are no exact data available regarding the amount of pesticides used in the Prespa region. Table 4 represents rough data on the use of pesticides in the entire Prespa region, calculations based on average quantities of pesticides used for one hectare of apple orchards per year and wheat production fields.

Pesticide type	Quantity (tons per year)	% of total
Fungicides	38.5	60%
Herbicides	3.2	5%
Insecticides	22.5	35%
Total	64.2	100%

Table 4. Use of pesticides in the Prespa region.

In total 64,000 kg of pesticides are used each year in the Prespa region. The behaviour of pesticides in the soil varies, some are easily soluble and move with water while others are less movable. It is hard to predict the movement of pesticides in general and each active matter together with other components used to produce pesticides should be investigated separately. It is obvious that much lower amount of pesticides is used in comparison with fertiliser use.

Due to the inconsistency of the current solid waste management system in the Municipality of Resen, including Golema River watershed, as well as the low public awareness, significant quantities of mainly organic (waste apples and yard waste), and partly hazardous (pesticide packaging) solid waste generated by the agricultural activity are being disposed of in the river channels and the riparian corridors. This inappropriately disposed waste has considerable negative impact on the surface, ground waters and soil, and especially on the Golema River, hence influencing the Prespa Lake system (Fig. 11).

4. Materials and methods

Aiming to resolve the detected intensive eutrophication of Prespa Lake watershed and to produce a management plan that is going to address this issue, the analyses presented in this chapter have actually been conducted in order to resolve the anthropogenic from natural processes of eutrophication. By focusing on the *reference conditions*, we have been able to detect the intensive human impact dated 1,500 years ago through massive deforestation and subsequent washout of nutrients into the Prespa Lake. This influence in combination with the very recent (only some 100 years) intensive pollution impacts in the watershed have triggered the turnover of the lake from the nitrogen towards phosphorus driven ecosystem and corresponding cyanobacterial toxic 'water blooms'.

Fig. 11. Waste apple dumping in Golema River and persistent foam on the Prespa Lake surface.

Assessment of the ecological status of the water bodies have to be based on the comparison of the level of deviation of current conditions of the biological quality elements in a water body with pristine conditions of the same one which means conditions without any anthropogenic disturbance. The level of deviation in terms of numeric Ecological Quality Ratios (EQR) have to be used for assigning an appropriate ecological class to a water body and describe its ecological status. Due to the usual practice of lack of continuous monitoring data on the biological quality elements which is necessary for description of the trend of the ecological condition of the aquatic ecosystems, the WFD leaves space for two approaches to be used for bridging this gap. A combination of the available historical data with at least a one year period of monthly detailed field monitoring of biological quality parameters gives a firm and reliable starting point in assessment and description of the ecological status of a certain water body. This approach combined with an iterative re-assessment process based on monitoring data of key type-specific biological quality elements to be derived during the river basin management plan implementation will enable a more comprehensive description of the ecological status of the water bodies and more efficient assessment of the effectiveness of the measures and activities implemented aimed at achieving the prescribed environmental objectives.

Bearing in mind the above, the development of the Prespa Lake Watershed Management Plan has been fully accompanied with identification, delineation and categorisation of the main surface and ground water bodies in the basin, and included establishment and implementation of a one year surveillance monitoring system of key biological quality elements with monthly field and laboratory analyses on the most important physical, chemical and biological parameters.

Development of a representative monitoring network of sampling sites in the Lake Prespa watershed included five lakes, four rivers and seven ground water (wells) sampling points. The most representative lake sampling points were chosen where continuous pressure from various anthropogenic activities was most expected. The river sampling points were the mouth waters of the four main rivers in the Prespa watershed flowing in the Prespa Lake. The ground water sampling points were chosen to represent the quantitative and chemical status of main ground water bodies, with the aim of providing clues as to the possible influence by the continuous pressure of various anthropogenic activities, as well as a certain correlation between the ground water bodies and surface water bodies. Figure 12 represents the chosen sampling points on the map.

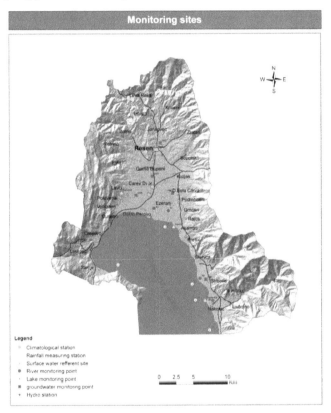

Fig. 12. Sampling points in Macro Prespa Lake watershed.

Five lake sampling sites were chosen as the most representative points for establishment of a relevant monitoring system of biological and physic-chemical quality parameters including: site Stenje (L1) village as a reference site for the deepest part of the lake in that area, a site in the vicinity of the mouth of Golema River (L2) for checking the impact of the river on the lake's ecosystem, a site in the vicinity of the mouth of the Kranska River (L3), a site in the vicinity of the mouth of the Brajcinska River (L4) and the Dolno Dupeni site (L5) as a reference site of the overall conditions at the border with Greece.

Four major river courses, Istocka, Golema, Kranska and Brajcinska (R1-R4), were permanently monitored during the 12 month sampling period at their mouths in Prespa Lake, in order to determine the overall qualitative and quantitative pressures on the lake's ecosystem. Their reference sites (R1(r)-R4(r)) were checked only once (March 2010) to establish the reference physic-chemical and biological conditions, and delineation of the water bodies.

Table 5 represents the full range of analysed parameters during the 12 month monitoring period and the frequencies used to monitor different environments or media.

Parameter	Rivers	Lakes	Sediments	Biota
Phytoplankton		Annually/12x		
Benthic macroinvertebrates	Seasonally/4x	Seasonally/4x		
Fish	Seasonally/2x	Seasonally/2x		
Phytobenthos	Annually/12x	Annually/12x		
Macrophytes	Seasonally/2x	Seasonally/2x		
Flow	Annually/12x			
Depth	Annually/12x	Annually/12x		
Temperature	Annually/12x (profiling)	Annually/12x (profiling)		
Transparency	Annually/12x	Annually/12x		
Suspended solids	Annually/12x	Annually/12x		
Dissolved oxygen	Annually/12x	Annually/12x (profiling)		
pH	Annually/12x	Annually/12x (profiling)		
Conductivity	Annually/12x	Annually/12x (profiling)		
Alkalinity	Annually/12x	Annually/12x		
Ammonium (NH_4)	Annually/12x	Annually/12x		
Nitrate (NO_3)	Annually/12x	Annually/12x		
Nitrite (NO_2)	Annually/12x	Annually/12x		
Total Nitrogen	Annually/12x	Annually/12x	Seasonally/4x	
Inorganic nitrogen	Annually/12x	Annually/12x		
Organic Nitrogen	Annually/12x	Annually/12x		
Ortho-phosphate (PO_4)	Annually/12x	Annually/12x		
Total phosphorus	Annually/12x	Annually/12x	Seasonally/4x	
Sulfate (SO_4)	Annually/12x	Annually/12x		
Calcium (Ca)	Annually/12x	Annually/12x		
Magnesium (Mg)	Annually/12x	Annually/12x		
Chloride	Annually/12x	Annually/12x		

Parameter	Rivers	Lakes	Sediments	Biota
COD	Annually/12x	Annually/12x		
BOD	Annually/12x	Annually/12x		
Cadmium (Cd)	Seasonally/4x	Seasonally/4x	Seasonally/4x	Seasonally/4x
Lead (Pb)	Seasonally/4x	Seasonally/4x	Seasonally/4x	Seasonally/4x
Mercury (Hg)	Seasonally/4x	Seasonally/4x	Seasonally/4x	Seasonally/4x
Nickel (Ni)	Seasonally/4x	Seasonally/4x	Seasonally/4x	Seasonally/4x
Arsenic (As)	Seasonally/4x	Seasonally/4x	Seasonally/4x	Seasonally/4x
Cooper (Cu)	Seasonally/4x	Seasonally/4x	Seasonally/4x	Seasonally/4x
Chromium (Cr)	Seasonally/4x	Seasonally/4x	Seasonally/4x	Seasonally/4x
Zink (Zn)	Seasonally/4x	Seasonally/4x	Seasonally/4x	Seasonally/4x
Iron (Fe)	Seasonally/4x	Seasonally/4x	Seasonally/4x	
Pentachlorobenzene	Seasonally/4x	Seasonally/4x	Seasonally/4x	Seasonally/4x
Hexachlorobenzene	Seasonally/4x	Seasonally/4x	Seasonally/4x	Seasonally/4x
DEPH	Seasonally/4x	Seasonally/4x	Seasonally/4x	
Nonylphenols	Seasonally/4x	Seasonally/4x	Seasonally/4x	
4-tert-Octylphenol	Seasonally/4x	Seasonally/4x	Seasonally/4x	
Naphtalene	Seasonally/4x	Seasonally/4x	Seasonally/4x	
Floranthene	Seasonally/4x	Seasonally/4x	Seasonally/4x	
DDT (6,7)	Seasonally/4x	Seasonally/4x	Seasonally/4x	Seasonally/4x
DDD (6,7)	Seasonally/4x	Seasonally/4x	Seasonally/4x	Seasonally/4x
DDE (6,7)	Seasonally/4x	Seasonally/4x	Seasonally/4x	Seasonally/4x
Total Microcystins (cell and free)		Annually/12x	Seasonally/4x	Seasonally/4x

Table 5. Parameters and frequencies analysed during the 12 month monitoring period in Macro Prespa Lake watershed.

4.1 Field sampling

Conducting analysis of physic-chemical parameters, priority substances, phytoplankton and phytobenthos quality parameters require a representative water and biological material sample to be collected, preserved and prepared for laboratory analyses. For consistent collection of materials a sampling manual is prepared. After collection, materials were transported to the Laboratory for Ecology of Algae and Hydrobiology at the Institute of Biology Faculty of Natural Sciences in Skopje for further treatment.

4.2 Laboratory analyses

Table 6 describes the analysed parameters, type of the analyses and the methods used during the 12 month surveillance monitoring in Prespa Lake watershed.

The spectroscopic analyses were conducted on Photometer System Max Direct-LoviBond® (www.tintometar.com) with the use of LoviBond® analytical kits with appropriate analytical range.

Parameter	Type of analysis	Type of method
Flow	Field Measurement	Hydrometrical wing
Depth	Field Measurement	Van Veen bottom sampler
Temperature	Field Measurement	Tintometar Senso Direct 150
Transparency	Field Measurement	Secchi Disk
Suspended solids	Laboratory Analysis	Analytical
Dissolved oxygen	Laboratory Analysis	Oxy meter - Tintometar Senso Direct 150
pH	Field Measurement	pH meter - Tintometar Senso Direct 150
Conductivity	Field Measurement	Conductivity-meter - Tintometar Senso Direct 150
Alkalinity	Laboratory Analysis	Analytical
Ammonium (NH_4)	Laboratory Analysis	Analytical-spectroscopic
Nitrate (NO_3)	Laboratory Analysis	Analytical-spectroscopic
Nitrite (NO_2)	Laboratory Analysis	Analytical-spectroscopic
Total Nitrogen	Laboratory Analysis	Analytical-spectroscopic
Inorganic nitrogen	Laboratory Analysis	Analytical-spectroscopic
Organic Nitrogen	Laboratory Analysis	Analytical-spectroscopic
Ortho-phosphate (PO_4)	Laboratory Analysis	Analytical-spectroscopic
Total phosphorus	Laboratory Analysis	Analytical-spectroscopic
Sulfate (SO_4)	Laboratory Analysis	Analytical-spectroscopic
Calcium (Ca)	Laboratory Analysis	Analytical-spectroscopic
Chloride	Laboratory Analysis	Analytical-spectroscopic
COD	Laboratory Analysis	Analytical-spectroscopic
BOD	Laboratory Analysis	Analytical-spectroscopic
Magnesium (Mg)	Laboratory Analysis	Analytical-AAS[2]
Cadmium (Cd)	Laboratory Analysis	Analytical-AAS
Lead (Pb)	Laboratory Analysis	Analytical-AAS
Mercury (Hg)	Laboratory Analysis	Analytical-AAS
Nickel (Ni)	Laboratory Analysis	Analytical-AAS
Arsenic (As)	Laboratory Analysis	Analytical-AAS
Cooper (Cu)	Laboratory Analysis	Analytical-AAS
Chromium (Cr)	Laboratory Analysis	Analytical-AAS
Zink (Zn)	Laboratory Analysis	Analytical-AAS
Iron (Fe)	Laboratory Analysis	Analytical-AAS
Pentachlorobenzene	Laboratory Analysis	Analytical-GC[3]
Hexachlorobenzene	Laboratory Analysis	Analytical-GC
DEPH	Laboratory Analysis	Analytical-GC
Nonylphenols	Laboratory Analysis	Analytical-GC
4-tert-Octylphenol	Laboratory Analysis	Analytical-GC
Naphtalene	Laboratory Analysis	Analytical-GC

[2] Atomic Absorption Spectrophotometry
[3] Gas Chromatography (GC)

Parameter	Type of analysis	Type of method
Floranthene	Laboratory Analysis	Analytical-GC
DDT (6,7)	Laboratory Analysis	Analytical-GC
DDD (6,7)	Laboratory Analysis	Analytical-GC
DDE (6,7)	Laboratory Analysis	Analytical-GC
Total Microcystins (Cell bound and extracellular)	Laboratory Analysis	Immunological-ELISA[4]

Table 6. Parameters, analyses and methods used during the 12 month surveillance monitoring on chosen sampling sites in Prespa Lake watershed.

The microcystin analyses were conducted with ELISA M956 Microplate Reader Metertech® with the use of Abraxis® Microcystins-DM ELISA 96 Microtiter Plates, product no: 522015 (www.abraxis.com). Analyses of microcystins in sediments used the same method with prior procedure of digestion and pre-treatment of the collected sediments. Analyses of microcystins in biological tissues were conducted by means of their extraction with appropriate procedure and subsequent ELISA analyses.

Heavy metals' analyses were conducted with Atomic Absorption Spectrophotometer Varian 10BQ.

Analyses for priority substances were conducted on Gas Chromatograph Varian.

Determination of chlorophyll pigments was conducted by the trichromatic method (Strickland and Parson, 1968) and measured on Photometer System Max Direct-LoviBond®.

Assessment of phytoplankton and phytobenthos species' composition was performed with microscopic analyses of the collected native and preserved algal material. The microscopic analyses of the collected material were performed on a light microscope Nikon E-800 with a Nikon Coolpix 4500 digital camera.

4.3 Hydrology

The quantities of the waters in the watershed were followed and measured by permanent water quantity stations and occasional (once per month) measurements of the water levels and flows, using the obtained rating curves $Q=f(H)$.

On each watercourse where permanent monitoring is not yet established, direct hydrometrical measurements were performed, in parallel with the water quality samplings, in order to determine the overall flux of pollutants into the lake ecosystem. The hydrometrical measurements were performed according to ISO standards: measurements' ranges are based on ISO 748:1979 and ISO2537:1988, using a hydrometrical wing for the flow measurements. Working methodology was undertaken according to the WMO's "Guide to Hydrological Aspects".

Water velocities measurements were performed by the usual methodology on several verticals in a perpendicular transect. The number of verticals should be uneven depending on the profile width, such as:

---

[4] Enzyme-Linked Immuno Sorbent Assay (ELISA)

- For profile of 10 m   3 to 5 verticals
- For profile of 10 to 50 m   5 to 7 verticals
- For profile of 50 to 100 m   7 to 9 verticals
- For profile of more than 100 m   9 to 15 verticals

The number of measuring points for water velocity on every vertical is dependent on the depth of that vertical, such as:

- On verticals up to h≤30 cm depth   one point on 0.5 h
- On verticals from h≤30 to >30 <100 cm   three points: 0.2; 0.6 and 0,8h
- On verticals from h≤30 and the bottom   in four points: 0.2; 0.6; 0.8 h
                                                              and bottom

4.4 Hydrogeology

Monitoring of the ground water level (GWL) of the ground water facilities (boreholes, wells) was executed occasionally (once per month) in the 12 month period (from March 2010 to February 2011). The measuring was performed using an electrical water level meter (type GTS, made in Italy) equipped with lightening and sonorous signal. In the case of the presence of artesian ground water level (which is expected in Pl_3 sediments – GW6), the monitoring was executed by measuring the natural capacity [l/sec] of the well.

4.5 Macrophytes

Monitoring of macrophytes in Lake Prespa was performed in the period from April to May 2010 and from August to September 2010 on five locations along the Macedonian coastline: Stenje, Golema River, Krajnska River, Brajcinska River and Dolno Dupeni.

The macrophytes field survey methods which have been used in the biological monitoring of Lake Prespa were consistent with the Water Framework Directive: a) **characteristic zonation** - determination if all type-specific vegetation zones for this Lake; b) **preparing the lists of species present in a lake (floristic inventories)** - determination of collected plant species using different floras and keys; c) **determination of vegetation density** - which focuses on the overall abundance of macrophytes. Abundance were estimated on a five-point scale (Braun-Blanquet 1964) where 1 = very rare, 2 = infrequent, 3 = common, 4 = frequent and 5 = abundant. The vegetation density is the result of different pressures, such as alteration of the shoreline, artificial water level fluctuations, artificial wave action and the trophic state; d) **mapping of lake's vegetation** (phyto-littoral mapping) - according to this method, the qualitative composition and distribution of aquatic vegetation was investigated.

4.6 Fish

Fish for analyses were collected with various fishing gear in day and night time experimental fishing. A cast net was also used for day time fishing with mesh size of 13 mm, whereas the night time fishing was performed with bleak nets (mesh size from 12 mm and 13 mm), and barbell nets with mesh size of 22 mm, 24 mm, 26 mm and 28 mm. The height of each fishing gear was basically a hundred heights per each mesh and the length was about 50 metres. All nets were set between 3 and 6 pm, fished overnight and lifted between 8 and 10 am in order to ensure that the activity peaks of each fish species was included.

The location of each net was selected according to proposed sampling sites in the project and based upon the past experience of the investigator.

All captured fish were identified, and length and weight was measured. The age and sex composition of the fish population from Lake Prespa were performed with standards ichthyological methods. Analyses of the scales were used for determination of the age structure of the fish populations, as well as length and weight growth. The obtained results were statistically processed.

4.7 Benthic invertebrates

Benthic invertebrates from the Prespa Lake and its main tributaries were collected according to the requirements of the EU Water Framework Directive (WFD) 2000/60/EC.

From the Prespa Lake itself, the collection of bottom fauna samples was performed by several different devices: Ekman grab, sediment corer, triangle bottom dredge and hand net. Macroinvertebrate standard methods applicable to lakes were used (ISO 9391:1995 and ISO 7828:1985). Concerning to the main tributaries of Prespa Lake, benthic invertebrate samples were collected with a Surber sampler or hand-net following standard methodology for collection of bottom fauna (ISO 8265:1988 and ISO 7828:1985). For preservation of biological samples 70% ethyl-alcohol or 4% formaldehyde were used, samples properly labelled and additionally processed in the laboratory.

In the laboratory, the animals were flushed with tap water through a standard sieve (280 µm pore size). Material was divided by groups, for further determination, mainly to the lowest taxonomic level (genus/species). Generally, determination to the species level is recommended because the species level logically provides the most detailed and sound information about autecological demands of a certain animal species. Determination was performed using identification manuals. Based on WFD principles, data informing on the communities' taxonomic composition, abundance, diversity and sensitive taxa were taken into consideration, both for Prespa Lake and its tributaries. Biotic indices that are suitable for Prespa Lake and its watershed monitoring purposes were used.

5. Results of the performed surveillance monitoring in Prespa Lake watershed

According to the historical hydrology data and observations, there are four major tributaries in the Macedonian part of Macro Prespa Lake which can be considered as separate rivers of significance and should be subsequently subdivided into the Istočka, Golema, Kranska and Brajčinska Rivers. Although small in it watershed, Kurbinska River was also taken into consideration because of it relatively significant quantity of water.

All of the other temporal water courses have been proven to be torrent carriers which usually drain forested areas and have little significance in the overall water quality analysis; although they may have a significant role in the water balance of the lake at specific times, as well as in flood hazard management because of their character.

Considering the Prespa Lake itself, there is a littoral plateau of approximately 14-16 metres depth that completely surrounds the lake and two major depressions – one near the village of Stenje (approximately 47 metres depth) and the other in the vicinity of Agios Germanos in Greece (approximately 58 metres depth) that represent the true profundal of the ecosystem. It is therefore a logical solution to delineate the water bodies in the Macedonian part of the Prespa Lake according to sampling sites presented on Figure 12, and to consider the littoral and profundal part of the lake as a one water body.

This delineation of the surface water bodies in the Macedonian part of Macro Prespa Lake has been used as a basis for a 12 month surveillance monitoring programme conducted in the period April 2010 - June 2011.

5.1 Rivers

All rivers in Prespa Lake watershed belong to the same river type (type 1 – siliceous rivers of mid-altitude and size) in the one eco-region 6 (Hellenic Western Balkan). They are mostly mountain type rivers with steep slopes and a stony bottom, normally with well-developed riparian vegetation; only the small part of the watercourses prior to their mouth in Prespa Lake may be slow flowing and on sandy or muddy substrata. In combination with the very small river lengths (and also catchments - <100 km^2), the only important driving force that might impact the natural ecological conditions in the rivers are human activities.

Due to the lack of a continual monitoring system in the area, the short timeframe of the project and the need for obtaining the most reliable data relevant for the development of the Management Plan, delineated river water bodies in Prespa Lake watershed were monitored for the basic physical, chemical (including major nutrients, priority substances and heavy metals) and biological parameters. Out of the WFD proposed biological quality elements, the categorization of rivers is chosen to be based on phytobenthos and macroinvertebrates. The reference conditions are also based on these parameters.

Basic physical parameters

The basic physical parameters measured during the established monitoring system in the course of the project (Fig. 13) reflect the overall natural conditions of the selected river water bodies. The only obvious sign of human impact are the values for *conductivity* and *total dissolved solids* which are much higher in Golema River and Istočka River. Values for *pH* are also worth notifying since in July they were also significantly lower than expected in Golema Reka River water bodies (3.92-5.17 respectively), but also in Kurbinska River (3.9!). This finding may be attributed to some specific pollution impact at the moment of sampling, but also to some other origin and therefore deserve investigative monitoring in future.

The values for dissolved oxygen generally decreased in the water bodies under intensive human influence (Golema River), but were still in the realm of a good ecological status; Kurbinska River is again an exception since there is no major human impact (apart of the water abstraction) recorded for this water body, but its DO value was only 5.4 mg $O_2 \cdot L^{-1}$. Nevertheless, the performed monitoring period was far too limited to enable any firm conclusions on the oxygen dynamics or the underlying causes; it should be substantially extended in future.

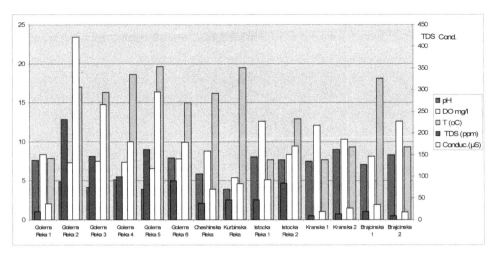

Fig. 13. Basic physical parameters of delineated rivers in Prespa Lake watershed.

Nutrient status

Measured major nutrients in delineated river water bodies in Prespa Lake watershed (Fig.14) were mostly based on *nitrogen* and *phosphorus* compounds, with the addition of *sulphates* as an indicator of the direct human influence. All of the examined water bodies show highly increased nutrient composition, especially regarding the *total N* and *P* compounds. In water bodies of the Golema and Istočka Rivers, *ammonia* and especially *sulphates* were very high, but their presence is also remarkable in the rest of the examined rivers. *Nitrites* were present in all examined water bodies, except in Kurbinska River, in low quantities, pointing to continual wastewater pressure. Detected high amounts of nutrients in examined river water bodies points to a complete lack of wastewater treatment, significant diffuse source pollution run-off and vast amounts of waste and agricultural and industrial discharge (refer to Fig. 15 for an estimation of the total annual load of nutrients by different river water bodies). These results might explain, in broad terms, the detected variations of *acidity* in examined waters which were most probably directly connected to the discharged nutrient quantities and the season of fertiliser application; all of the examined nutrients, especially sulphates, have high potential of forming acids which have rapid or prolonged impact on water biota and chemistry. In other words, river water bodies in Prespa Lake watershed have very low acid neutralising capacity caused by prolonged deposition of acidifying chemicals.

Amid the lack of categorisation of surface waters according to nutrients in the Macedonian legislation (only the values for *total N* and *ammonia* are considered, but with wrongly stated units of measurement; *Official Gazette of RM 18/99*), the detected values for *phosphates, sulphates, total N* and *ammonia* (Fig.14) place the lower parts of the examined water bodies highly in the V water quality class. In this respect, Golema River and Istočka River have the maximum detected levels, but even Kranska River and Brajčinska River have been shown to experience intensive pressure from nutrients.

According to results presented in Figure 15, *sulphates* are the major nutrient carried to Prespa Lake mostly by Brajčinska, Istočka and Golema Rivers. Significant amounts of *nitrogen* and *phosphorus* compounds are also recorded in all major water bodies, except Kurbinska River which has been found not to be subject to intensive human impacts regarding nutrients.

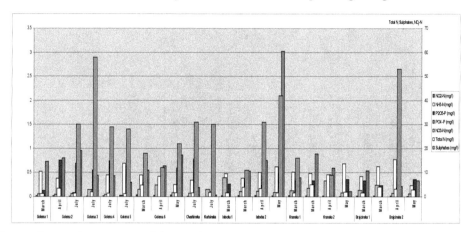

Fig. 14. Detected nutrients in delineated rivers in Prespa Lake watershed.

Conducted surveillance monitoring is far from sufficient to enable reliable quantification of the total nutrient pressure in the Prespa Lake system. Nevertheless, the obtained estimations of cumulative nutrient loads (Fig. 16) are in concordance with the estimations done for the diffuse source of pollution coming from agriculture. Furthermore, these results disprove the hitherto reported estimations (Grupce, 1997; Naumovski et al., 1997) of the phosphorus load in Prespa Lake of 84 tons per year (of which 41 tons per year are coming from natural processes and 43 tons per year are due to anthropogenic activities) by almost double amount (Fig. 16), without including all of the possible emissions (diffuse sources, natural run-off, etc.).

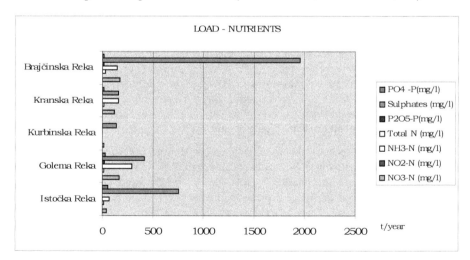

Fig. 15. Estimated total nutrient load detected in investigated rivers of Prespa Lake watershed.

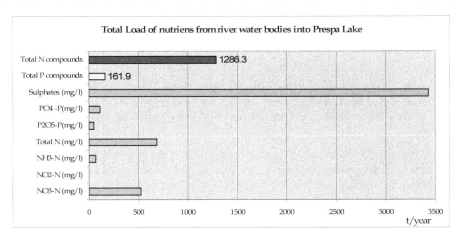

Fig. 16. Estimation of the cumulative load of nutrients coming from all river water bodies into Prespa Lake.

Heavy metals

Three heavy metals dominate the river water bodies of Prespa Lake watershed: *manganese, iron* and *aluminium*. Their recorded concentrations are usually way beyond the permissible levels for natural conditions (III-IV water quality class) and therefore denote an intensive human origin (Fig. 17). Again, Golema and Istočka Rivers express the highest concentration levels, while the lower segment of Kranska River recorded very high *manganese* and slightly lower *iron* concentrations. Upper segments of all river water bodies (the reference conditions) have been found with heavy metals concentrations several times lower than the lower parts, thus reflecting the natural background emissions to the water habitats.

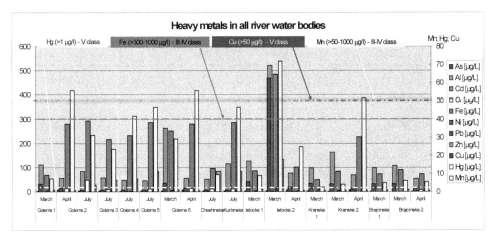

Fig. 17. Detected concentrations of heavy metals in river water bodies of Prespa Lake watershed.

Two other heavy metals, *copper* and *zinc*, have also been detected in significant quantities in the lower river parts of the Prespa Lake watershed, being most pronounced in the Istočka River. Their increased presence is the inevitable result of adverse human impact in the region.

The most severely influenced river water bodies, Golema and Istočka Rivers, are also characterised with a marked presence of *mercury, lead* and *arsenic*. Their toxicity and harmful effects on the environment and humans are well known and do not need any further elaboration.

Figure 18 gives an overview of the total intensity of the heavy metal load detected in the mouth waters of river water bodies prior to entering the Macro Prespa Lake. It is very clear that Golema and Istočka Rivers are the major source for all measured heavy metals, carrying more than 8 tons of *iron* and *aluminium* annually into the lake, but also significant amounts of *manganese, zinc, copper* and *lead* or *mercury*. Rivers Brajčinska and Kranska also add to the overall heavy metal load (Fig. 19), but to a much lesser extent.

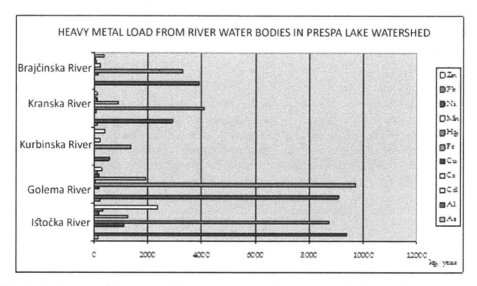

Fig. 18. Estimation of heavy metal load from river water bodies in the Prespa Lake watershed.

In summary, according to calculated estimations, Macro Prespa Lake receives more than 27 tons of *iron* and almost 26 tons of *aluminium* per year, coming from its major tributaries. It is also loaded with 4.6 tons of *manganese*, 3.5 tons of *zinc* and more than 1.5 tons of *copper* per year. Toxic metals are less abundant (563, 504, 132 and 118 kg per year for *arsenic, lead, chromium* and *mercury* respectively), but they do represent a significant load and a dangerous hazard to water biota (through processes of bioaccumulation) and humans. At this point, we do not have any information on the intensity of accumulation of the various heavy metals in water biota, specifically in fish; therefore these investigations will be proven invaluable for the proposed WFD monitoring system to be established in Prespa Lake watershed.

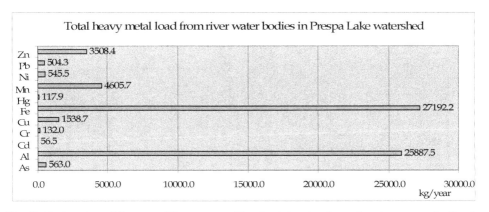

Fig. 19. Estimation of the overall heavy metal load originating from the river water bodies in Prespa Lake watershed.

Priority substances

The presented results for WFD priority substances are the first record of this kind for the Prespa Lake watershed (Fig. 20). Out of the proposed priority substances for the surveillance and operational monitoring purposes (WFD 2008), the comprehensive analyses performed so far in Prespa Lake watershed embrace *Chlorinated aromatic hydrocarbons, Poly-aromatic hydrocarbons (PAHs), Poly-chlorinated biphenyls (PCBs), Organophosphate pesticides, Phenols, Phthalates* and *Organochlorine pesticides*.

A total of 18 priority substances (out of the whole set of 80 analysed chemical compounds) were detected in river water bodies in Prespa Lake watershed (Fig. 16). *Bis(2-Ethylhexyl)phthalate* was present in almost all samples, the highest values recorded in Golema and Brajčinska Rivers. *Dibutilphthalate* was also found in all river water bodies, except Kurbinska River, but in slightly lower concentrations. *Organochlorine pesticides* were recorded in different concentrations and were dominant in investigated water bodies. *Gama-HCH (Lindan), Alpha HCH* and *Alpha Endosulfan* were the most commonly present in all water bodies, but the very high values for *Heptachlor* in Golema River 6 and especially in Kranska River. In summary, it is very clear that all of the examined river water bodies in Prespa Lake watershed are under prolonged influence of a significant pollution pressure coming from excess utilisation of various pesticides groups, plastic materials and industry. Even the uphill mountain river parts not subjected to any significant (visible) human impact, that should be used as reference conditions in the watershed, are also under obvious pressure; DDE and DDD although banned from utilisation are still present in Golema 1, Kranska 1 and Brajčinska 2 water bodies. These results point out that the surface water bodies in Prespa Lake watershed have been, and still are, subjected to intensive pressure coming from agriculture and irregular waste disposal. Apart from an *in situ* pollution of the surface waters, aero deposition might play a significant role especially in polluting the upstream mountain river segments with pesticides or their residues. Although in domestic legislation only *Heptachlor* and *PCB* are listed in water categorisation tables (both found in concentrations stated for V or III-IV category respectively), other detected substances will also have to be included in future.

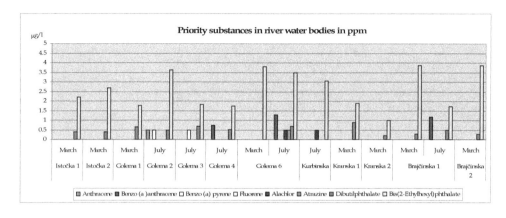

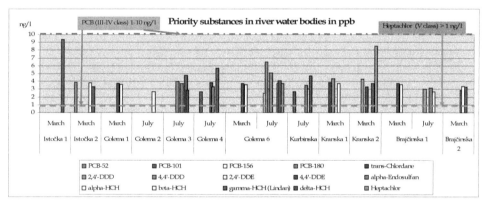

Fig. 20. Detected priority substances (without heavy metals) in river water bodies of Prespa Lake watershed.

The final conclusion on detected priority substances in the river water bodies in Prespa Lake watershed is that numerous different compounds were detected, some of them with very high concentrations (III-IV or V water quality class) and that all of them represent an elevated risk for the environment, water biota and humans. Toxic and already forbidden chemicals like *DDD* or *DDE* are still present in the waters what underlines their constant utilisation. Apart from detected pesticides or their derivatives, detected *phtalathes* as main chemicals in the production of plastics (Fig. 21) directly emphasise irregular solid waste treatment or disposal, and insufficient industrial and domestic wastewater treatment.

It is very critical to note that the spreading of detected priority substances also affects the remote regions not subject to any visible human influence in the Prespa Lake watershed thus endangering the environment and biota in already declared protected areas, like NP "Pelister" and "Galichica". Their potential for bioaccumulation and prolonged devastating impact should be the focus of the Management Plan and consequent reduction measures.

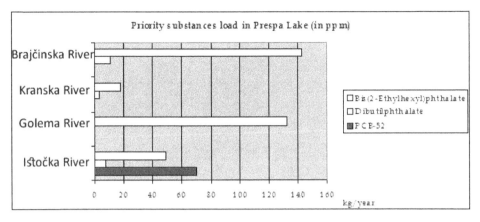

Fig. 21. Priority substances load to Prespa Lake coming from river water bodies in the watershed (in ppm).

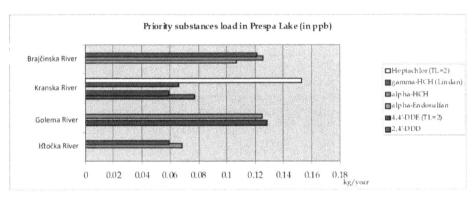

Fig. 22. Priority substances load to Prespa Lake coming from river water bodies in the watershed (in ppb).

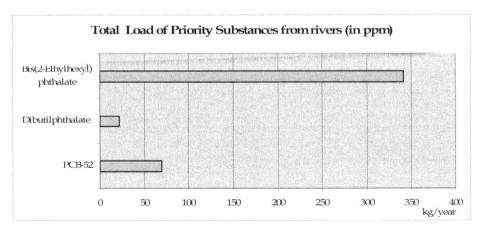

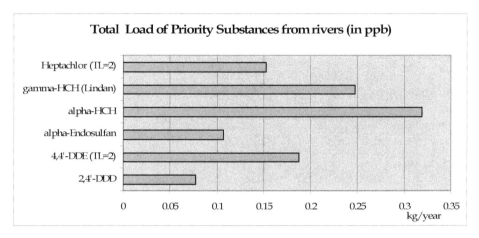

Fig. 23. Estimated total load of priority substances to Prespa Lake coming from river water bodies in the watershed (in ppm and ppb).

Ecological quality elements

Having in mind that there are practically no useful data (apart from the one project report – Krstic, 2007) on WFD ecological quality elements for the rivers in Prespa Lake watershed, the only feasible and promising approach was to use benthic organisms, algae and macrozoobenthos, to evaluate the ecological conditions in rivers. For the same reasons, there was no firm scientific opportunity to run any statistical calculations on the obtained data without significant bias and/or error.

The principal approach used in this situation was to determine the benthic algal and macrozoobenthic assemblages in upper river segments (river water bodies not subject to any significant human influence) to parts of the rivers where increased human interference in water physical or chemical properties was detected (usually lower river stretches after the point or diffuse wastewater input or leaching coming from cultivated land). Using this approach, we were able to detect the benthic assemblages in natural conditions (or reference conditions) for the rivers in Prespa Lake watershed, and also to reveal to the highest possible extent the succession of benthic communities under the detected human pressure. All of the methods for sampling, handling and analysing the samples were performed according to WFD recommended ISO or CEN standards (ISO, 2003; CEN, 2003).

Regarding the WFD proposal for using biotic indices in the process of describing the ecological status of water bodies, our approach was ruled by the following facts and postulates:

i.   The biotic indices in relation to water quality monitoring (for both algae and macrozoobenthos) have been developed in different countries based on long-term monitoring data and detected autecological preferences of different taxa. In the case of Prespa Lake and the water bodies in its watershed, neither of those lines of data are available. Even more so since there were numerous diatoms (a group of algae) described new to science in the region (Levkov et al., 2006) for which there are no data in the literature;

ii.   Using biotic indices developed for any organism in a region (country, continent) other than the region (river, lake, watershed) in question has already been proven erroneous on multiple occasions (Van Dam et al., 2005; Krstic et al., 2007). This is fundamentally based on differences in the *total capacity* of every ecosystem, numerous (not accountable) sources of variation of the ecological parameters and direct human influence. Since all of the indices are derived via assigning specific numbers to organisms (algae in particular) to be used in formulae calculations, the only scientifically sound approach is to develop a specific index for every water body;

iii. The last (and not least) reason against using bio-indices other than a basic community similarity (or dissimilarity) index is their gross oversimplification of the biological (or genetic) responses of living biota to different environmental conditions. The same organism (if we exclude taxonomical errors as yet another frequently very misleading source) cannot react in the same way in different environments (habitats); therefore, using and comparing biotic indices based on species variations (numbers in the population or different biological treats) between habitats seems (and has been proven) highly inapplicable.

With all the stated limitations in focus, class boundaries among different water quality classes and the reference conditions were based on benthic flora and fauna community structure in which dominant taxa characteristics, and therefore different classes, are recorded. In the case of benthic algae, clear distinction among class boundaries are based on the occurrence of taxa indicative for higher eutrophication or saprobity levels, or mass occurrence of specific cyanobacteria indicating very high pollution. Macrozoobenthos communities were subjected to evaluation according to the Danish Stream Fauna Index (DSFI – Skriver et al., 2000) and EPT richness (Bode et al., 1997), but we still express our reservations regarding the above mentioned restrictions for application of the biological indices. To achieve the best possible level of certainty regarding the biological quality elements in Prespa Lake watershed, the surveillance and investigative monitoring must be significantly extended.

In general, algal assemblages in river water bodies of Prespa Lake watershed are quite distinctly separated along the eutrophication or pollution gradient detected in examined watercourses. Namely, the upper river segments of all rivers are dominated by diatom flora characteristics for clean, slightly acidophilic, oligotrophic waters or **good ecological conditions** composed of *Diatoma hyemalis, Diatoma mesodon, Hannea arcus, Meridion circulare, Meridion circulare var. constricta, Eunotia minor, Achnanthidium jackii, Docuooata hexagonu, Encyonema mestanum, Krsticiella ohridana, Pinnularia eifelana, Pinnularia sudetica, Psammothidium daonense*. The rocky bottom of these river parts is usually not covered by any visible flora, but occasionally fragments of water mosses, typical chrysophyte (golden algae) *Hydrurus foetidus* colonies or the red alga *Lemanea fluviatilis* which are usually deprived of any rich epiphytic diatom flora. Therefore, this algae assemblage is chosen to be the indicator for the *reference conditions of all rivers* in Prespa Lake watershed.

The second type of algal assemblage is found in lower river stretches prior to or after the major pollution events indicating **moderate ecological conditions**; Golema Reka 2, 3, 4, Češinska, Kurbinska, Kranska 2 and Brajčinska 2. This assemblage is clearly dominated by diatom taxa indicative for higher nutrient concentrations in the water and more intensive decomposition processes, such as: *Melosira varians, Fragilaria capucina, Ulnaria ulna, Achnanthidium lanceolatum, Cocconeis placentula* var. *euglypta, Navicula phyllepta, Navicula cryptotenella, Navicula halophila, Navicula lanceolata, Navicula tripunctata, Navicula cryptocephala, Frustulia vulgaris, Reimeria sinuata, Gomphonema olivaceum, Gomphonema angustatum, Gomphonema micropus, Gomphonema aff. olivaceoides, Encyonema silesiacum, Amphora pediculus, Nitzschia linearis, Nitzschia palea, Surirella minuta*. Usually, the green branched alga *Cladophora glomerata* is also markedly present in these water bodies, bearing a rich epiphytic growth usually by *Cocconeis placentula* var. *euglypta*, but sometimes blue-green cyanobacteria *Heteroleiblenia kossinskajae* or *Pseudoanabaena limnetica* were observed as significant epiphytes.

Finally, the last detected algal assemblage was found in the most severely polluted river bodies of Prespa Lake watershed, like Golema Reka 5, 6, 7, and Istočka 2, and thus denoting **bad or poor ecological conditions**. The typical example of this algal assemblage was found at Golema Reka 5 sampling site where the diatom flora is reduced to as much as only 3 - 4 taxa, like *Nitzschia palea, Navicula cryptotenella* and *Ulnaria ulna* representing more than 98% of all detected cells. Highly decreased algal biodiversity is replaced by a mass development of two cyanobacterial species *Pseudoanabaena limnetica* and *Phormidium limosum* which completely cover the rocks on the bottom.

The WFD requires classification, in terms of ecological status, for all European surface waters. The classification should be based on reference conditions, which are intended to represent minimal anthropogenic impact and observed deviation from these conditions (Andersen et al., 2004). For each surface water body type, type-specific biological reference conditions were established, representing the values of the biological quality elements for that surface water body type at high ecological status.

Among the biological communities, the macrozoobenthos is by far the most frequently used bioindicator group in standard water management (Hering et al., 2004). Numerous biotic index and score systems have used macrozoobenthos in the assessment of running waters (Rosenberg & Resh, 1993). The most represented biotic index or score methods are: taxa richness, number of EPT taxa, Saprobic Index (SI), Biological Monitoring Working Party (BMWP) Score, Average Score Per Taxon (ASPT), Danish Stream Fauna Index (DSFI). All indices were part of the respective national method planned for biological monitoring in the context of the Water Framework Directive (Birk & Hering, 2006).

Thus, in the frame of the project "Development of Prespa Lake Watershed Management Plan" and according to WFD requirements, categorisation of the delineated water bodies in the Prespa Lake watershed based on macrozoobenthos was done. Two metrics (EPT richness and DSFI) in assessment of the ecological health of the rivers were used. These metrics were selected because statistical power to detect a difference between the nutrient enriched and non-impacted sites was >0.99 for total taxa richness and number of EPT taxa. DSFI also had relatively high power >0.95 (Sandin & Johnson, 2000) (Table 7).

In order to assess the ecological conditions of the river water bodies and related macrozoobenthos assemblages, the analyses of collected samples were performed to detect the presence of so called *positive* (like Ephemeroptera, Plecoptera, Trichoptera, Diptera, Gammaridae and even Astacidae) versus *negative* taxa (usually Oligochaeta – Chironomidae or Tubificidae). Pollution sensitive taxa like *Ecdyonurus venosus, Baetis alpinus, Capnia vidua,*

Brachyptera risi and *Potamophylax latipennis, Crenobia alpina* and even crayfish *Austropotamobius torrentium* were compared for the percentage abundance within the community, and used to derive the EPT and DSFI indices.

EPT richness	DSFI value	Water quality
>10	7	high (reference condition)
6-10	6	good
2-5	5	moderate
<2	4	poor
	1-3	bad

Table 7. Water classification based on EPT and DSFI.

5.2 Prespa Lake

Out of the WFD's biology quality elements, phytoplankton, zoobenthos, macrophytes and fish were in the focus of investigations, supported by a full range of physic-chemical analyses including heavy metals and priority substances. Water for the chemical analyses was sampled as a collective sample from the full water column on the site or as sediment, while the basic physical parameters were measured at every sample depth. Special attention was paid to the sampling of plankton for algae and benthic habitats (littoral, sub-littoral and profundal) for the macrozoobenthos analyses.

In order to determine the *reference conditions* for Prespa Lake, analyses of core samples dated 10 ka (500, 1,000, 2,000, 5,000 and 10,000 years respectively; the deepest analysed core sample from approximately 30 metres of the core depth) were performed for the first time regarding the total phosphorus content and diatom composition.

Macrophytes and fish samples from the selected sampling sites on Prespa Lake were collected during June 2010 and according to WFD sampling guides.

Basic physical parameters

The basic physical parameters detected in the waters of Prespa Lake (Fig.24) revealed some interesting features of this unique ecosystem. For example, recorded temperatures show a normal and gradual increase towards warmer months, but there was no sharp and rapid decrease in one water layer (thermocline) during the warmest month (July 2010) although the water temperatures between the deepest and the shallowest parts differ by more than 10°C. This may be a result of a very turbulent climate in the sampling period with constant mixing of the water layers or as a consequence of intensive discharge of the deep water sources (sub-lacustrine water sources) again related to the rainy season. High deep water temperatures of 14-15°C also indicate the full intensity of thermal insulation and possible full scale mixing of the entire water column during storms, which results in a constant nutrient supply of the epilimnion layer.

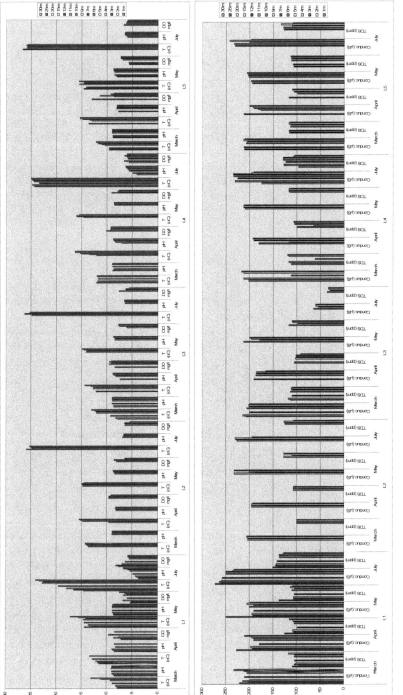

Fig. 24. Basic physical parameters detected at Prespa Lake sampling sites. The above findings are additionally corroborated by *dissolved oxygen* results. Deep water layers had showed relatively high DO values in the summer months, their values even being higher than upper water layers. Maximal DO values were recorded at around 3–4 m depth layers where the phytoplankton was in maximum development. These results are sharply opposite to frequent statements of oxygen depletion in the deep water layers of Prespa Lake; the lowest recorded values during our investigations were 5–5.3 mg·L^{-1} at the deepest part of the L1 sampling site in May 2010.

Another peculiarity recorded in these investigations was a very low pH reaction of the deep waters (beyond 10 m on L1 and L4) in July 2010, ranging as low as 3.7! This condition can arise if decomposition (bacterial) of organic material releases carbon dioxide and thus increases the amount of dissolved carbon dioxide; an increase in carbon dioxide decreases pH. Organic acids, often expressed as dissolved organic carbon (DOC), also decrease pH (Strumm & Morgan, 1981). Both of these possibilities point to the increased pressure on the lake in the spring-early summer period, which was also recorded in the data for the rivers in Prespa Lake watershed (Fig. 13).

Conductivity and TDS results obtained for the Prespa Lake (Fig. 20) sampling sites are in the realm of natural conditions for this type of lake (for example, comparison to Dojran Lake which shows conductivity values at around 800 – Krstić, 2011). Nevertheless, there is a slight increase of conductivity and corresponding TDS values in the deep water layers in July 2010, thus supporting the above statements of release of charged ions (either by microbial activity or human input) in this period.

Nutrient status

Detected nutrient levels in Prespa Lake sampling sites (Fig. 24) fully reflect the overall conditions already established for the watershed. The lake is dominated by sulphates, the same as the rivers in the watershed (Fig.14), but there is also a marked presence of total N basically due to elevated concentrations of nitrates and ammonia. Regarding ammonia, the whole investigated area was found to be in the III-IV category class as stated in the domestic legislation, while with regard to the total presence of nitrates Prespa Lake has to be declared as a Nitrate Vulnerable Zone as stated in EU legislation[5].

With respect to *phosphorus* content in the waters of Prespa Lake the situation is even worse; detected *total phosphorus concentrations* (based on the sum of detected values for P_2O_5-P and PO_4-P) place the lake in realm of hyper-eutrophic conditions, both regarding domestic and EU legislation.

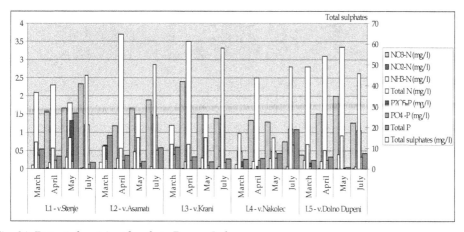

Fig. 24. Detected nutrient levels in Prespa Lake.

[5] Nitrates Directive (91/676/EEC).

Heavy metals

Contrary to presented results for the river water bodies (Fig. 13), the concentrations of heavy metals detected in the waters of selected Prespa Lake sampling sites (Fig.25) point to *copper, iron* and *zinc* as dominant metals. For *copper* the detected values were almost entirely in III-IV category, but the increased presence of the other two metals in detected concentrations also confirms the prolonged input.

There is also a marked presence of the toxic *mercury* and *arsenic* in Prespa Lake waters. Their increased detected concentrations in July 2010 clearly support the argument for intensified human activities in the spring period; *mercury* appears at the L2 sampling site as a result of the Golema River influence and is detected almost in all samples with concentrations high above the V water quality class. Its presence is striking on L5 (v.Dolno Dupeni) with almost eight times increased concentration compared to the 1 mg·L⁻¹ level for V quality class. *Arsenic* is also present in all sampled waters of Prespa Lake, but in much lower concentrations than *mercury*. It rose to III-IV water quality range only in L4 (v.Nakolec – mouth waters of river Brajčinska) in July 2010. Nevertheless, its accumulation and persistence in Prespa Lake waters is evident.

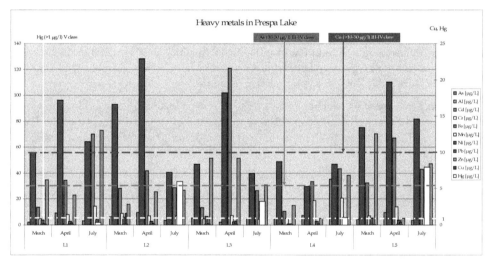

Fig. 25. Detected concentrations of heavy metals in the water of selected sampling sites of Prespa Lake.

Priority substances

Prespa Lake waters were also found to contain 20 priority substances out of more than 70 analysed during March and July, 2010. The same applied for the river water bodies, with *Bis(2-Ethylhexyl)phthalate* and *Dibutilphthalate* dominating in the samples of Prespa Lake waters. But, there was also a marked presence of *Benzo (a) pyrene, Benzo (a) anthracene* and *Naphthalene* (Fig. 26).

Considering the presence of pesticides or their residues in the Prespa Lake waters (Fig.27), *Gama HCH (or Lindan)* is dominantly found on all sampling sites followed by *Heptachlor* which is usually in concentrations high above the permissible 1 ng·L^{-1}. It is interesting to notice that the L2 sampling site in the vicinity of the mouth waters of Golema River was found with the lowest number of detected priority substances, contrary to the expected and detected pressures coming from this water body. The other sampling sites along the North-East coast of the lake (L3-L5) and the deepest part on L1 sampling site had a much higher number of detected priority substances and maximal values of separate chemicals. These findings corroborate the proposed intensive mixing of the Prespa Lake waters with significant underwater currents that are spreading the pollution impact to a much wider area.

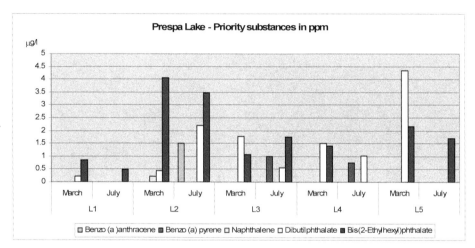

Fig. 26. Priority substances (in ppm) detected in waters of selected Prespa Lake sampling sites.

Priority substances detected in Prespa Lake pose a significant hazard to biota and humans.

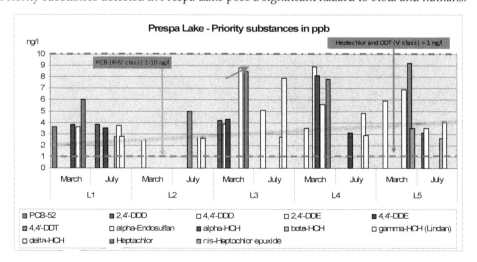

Fig. 27. Priority substances (in ppb) detected in the waters of selected Prespa Lake sampling sites.

By comparing the obtained results on priority substances for the river water bodies and Prespa Lake sampling sites (Fig. 28 and 29) interesting correlations could be formulated. Substances detected in high concentrations in the rivers, like *Bis(2-Ethylhexyl)phthalate* or *gamma-HCH (Lindan)* remain high in the lake's waters as well. Although others that were not recorded in very high concentrations in rivers, such as *Dibutilphthalate* or *Heptachlor*, show much higher concentrations in the lake, while *PCB's* tend to disappear from the lake's waters. These findings shed light on the very complicated and unpredictable pathways the detected priority substances have in the Prespa Lake ecosystem and point to the fundamental necessity to monitor and reveal their final destiny and impact they pose to the ecosystem, biota and human health.

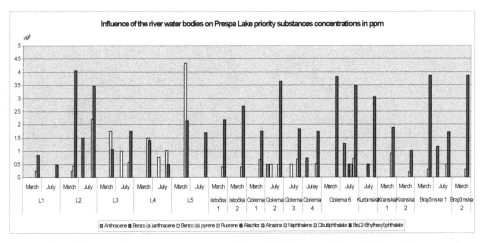

Fig. 28. The influence of river ecosystems on Prespa Lake regarding priority substances (in ppm).

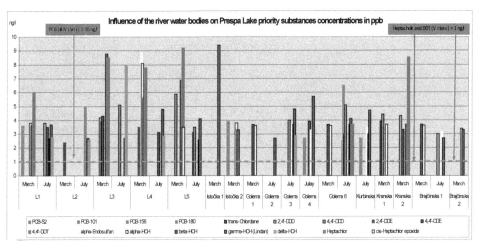

Fig. 29. The influence of river ecosystems on Prespa Lake regarding priority substances (in ppb).

The final analysis of the priority substances in Prespa Lake watershed was performed on sediment samples from the selected sampling sites in the lake (Fig. 30). At this point, only *gamma-HCH (Lindan)* was detected in the sediments of all sampling sites while the results for the L4 sampling site revealed sedimentation of the greatest number of analysed substances. Further monitoring of priority substances in Prespa Lake watershed should include far more frequent samplings and other media (like biota) in order to obtain overall conclusions about their patterns and role in the ecosystem.

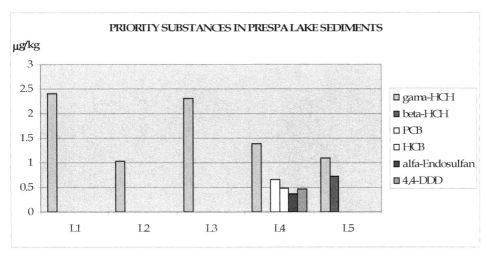

Fig. 30. Results on priority substances obtained by analysis of sediments in Prespa Lake sampling sites.

Ecological quality elements

Investigations of biological quality elements in selected Prespa Lake sampling sites (L1-L5) were performed on all selected WFD's Ecological Quality Elements[6] - *phytoplankton, phytobenthos, invertebrate fauna, macrophytes* and *fish*.

Benthic diatom communities in Prespa Lake have been recently well documented (Levkov et al., 2006) and again confirmed with the performed investigations. Nevertheless, their ecological preferences are more elusive since very limited investigations in that context have been performed so far on the lake. The greatest difference in diatom composition between Eastern and Western coast of Prespa Lake was observed in the littoral zone. The bottom of the West coast (from Stenje village to Perovo village) is covered with organic sediment. Beside dominant species *Cavinula scutelloides, Navicula rotunda, N. subrotundata* and *Amphora pediculus*, characteristic species from genera *Aneumastus* and *Sellaphora* are frequent in the benthic communities. The Eastern coast (D. Dupeni village to Pretor) is mainly covered by sand or a mixture of sand and organic sediment. In this region, several *Navicula* sensu stricto taxa are sub-dominant. It is supposed that distribution of these species is influenced by substrate. A few species so far known only from Lake Prespa could be found on both sides.

---

[6] WFD-Guidance document No.7: Monitoring under the WFD, 2003.

Diatom assemblages on larger depths are dominated by *C. ocellata* complex, as well as species with heavily silicified valves as *Diploneis mauleri*, *Campylodiscus noricus*, *Cymatopleura elliptica* and *Navicula hasta*.

Phytoplankton in Prespa Lake is much more uniform. It is usually composed of planktonic diatoms, like the *Cyclotella ocellata* complex, with very rare presence of algae belonging to other taxonomic groups, like the chrysophyte *Dinobryon bavaricum* during the winter months (November-April). But, there is a rapid change in the dominance during the summer months when strong development (more than 90% of dominance) of potentially toxic cyanobacteria like *Anabaena* sp. or *Aphanizomenon* sp. are usually observed, and clearly documented during the latest investigations in the frame of this project (Fig. 31).

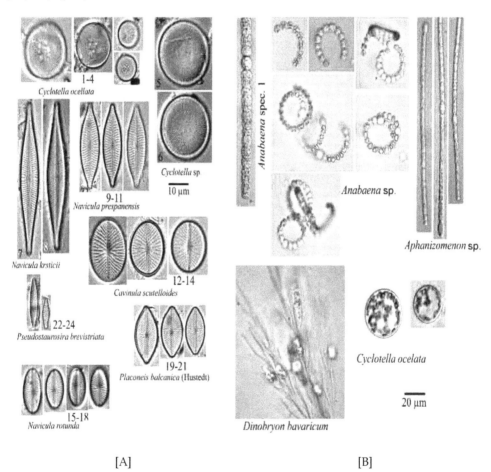

[A]   [B]

Fig. 31. Dominant algae in Prespa Lake. [A] Benthic diatoms, [B] Dominant blue-green cyanobacteria, chrysophyte *Dinobryon bavaricum* colony and a diatom *Cyclotella ocellata* in plankton.

In contrast to the extensive use of benthic invertebrates in river monitoring, ecological assessment in lakes has instead focused mainly on the response of open-water phytoplankton (usually measured as concentrations of chlorophyll a) to nutrient (mainly phosphorus) enrichment (OECD, 1982) and, to a lesser extent, that of profundal or sub-littoral communities (Dinsmore et al., 1993). While the WFD has proposed a need for monitoring of littoral communities in lakes, their use in regional monitoring of lakes in Europe has been very limited. The lack of incorporation of littoral invertebrates into lake monitoring programmes reflects a traditional and common view that the structural heterogeneity of lake littoral areas, and associated variable distribution of benthic macroinvertebrates, negates the feasibility of their use in ecological assessment (Downes et al., 1993). Many anthropogenic impacts that affect rivers (Boon, 1992) however, also affect lakes, and would be expected to drive changes in the littoral macroinvertebrate community.

In the frame of this project, according to WFD requirements, categorisation of the Macro Prespa Lake based on macrozoobenthos was done. Macroinvertebrates from five sampling sites and different depth regions were collected. Detailed analyses on composition, abundance, diversity of benthic invertebrate fauna and relative contribution of sensitive and tolerant invertebrate taxa were performed. Based on the recommendation by the Swedish Environmental Protection Agency, two indices of bottom fauna were used for ecological assessment of Prespa Lake: the Benthic Quality Index (BQI) based on profundal fauna and the Shannon Diversity index (H') based on littoral fauna. The corresponding water classification is given in Table 8.

Benthic Quality Index – BQI (Profundal fauna)	Shannon Diversity Index – H' (Littoral fauna)	Water quality
>4	>3.00	high
3-4	2.33-3.00	good
2-3	1.65-2.33	moderate
1-2	0.97-1.65	poor
<1	<0.97	bad

Table 8. Water classification based on BQI and H' for bottom fauna.

In Lake Prespa, the macrophyte vegetation shows relatively high species diversity in different parts of the littoral region. A high number of species is recorded at localities Golema River – 24, Asamati - 23 and v. Stenje – 21, while macrophyte species number is quite lower at Brajcino - 13 and Dolno Dupeni - 12. Recorded differences in the number of macrophyte species are most probably a result of different ecological conditions present at investigated localities, especially regarding nutrients. Namely, the presence of a higher species number at localities Golema River, Asamati and v.Stenje implies a very intensive anthropogenic influence. These areas of the littoral region have an increased presence of organic and inorganic material, what enables intensive growth and development of more diverse macrophyte vegetation.

Due to a decrease of the Prespa Lake water level in the last decade, marsh vegetation (reed and other emerged plants) progressively expanded around the lake (previously submerged littoral). Obtained results show that the dominant emerged plant in all investigated localities was *Phragmites australis* (with a density of 5 according to five-point scale). The reed forms a natural discontinuous belt around the lake comprised of numerous dense complexes. Other representatives of emergent vegetation were present in the belt of reed and in particular localities, where they formed almost pure associations. *Phalaris arundinacea, Typha latifolia, Typha angustifilia* were present with a density of 2, while *Shoenoplectus lacustris, Scirpus sylvaticus, Heleocharis pallustris* and *Cyperus longus* were present with a density of 1. Nevertheless, to obtain detailed information about changes in composition and spatial disposition of the vegetation, aquatic and marsh vegetation, long-term investigations are needed.

The fish population of the Prespa Lake is composed of 23 species of which 11 are autochthonous: *Alburnoides prespensis, Alburnus belvica, Anguilla anguilla, Barbus prespensis, Chondrostoma prespense, Cobitis meridionalis, Cyprinis carpio, Pelasgus prespensis, Rutilus prespensis, Salmo peristericus* and *Squalius prespensis*.

In the previous period 12 allochthonous species were introduced in Lake Prespa: *Carassius gibelio, Ctenopharyngodon idella, Gambusia holbrooki, Hypophthalmichthys molitrix, Lepomis gibbosus, Oncorhynchus mykiss, Parabramis pekinensis, Pseudorasbora parva, Rhodeus amarus, Salmo letnica, Silurus glanis* and *Tinca tinca*.

Alburnus belvica and *Rutilus prespensis* were caught in the greatest number of specimens at all investigated localities from Lake Prespa. On the contrary, *Anguilla anguilla, Chondrostoma prespense, Cobitis meridionalis, Pelasgus prespensis* and *Salmo peristericus* were not caught at any of investigated localities. Also, in the catches from all investigated localities *Cyprinis carpio* was present, but in very low numbers.

Populations of some of the introduced species are so reduced that they are very rarely present in representative and experimental fishing, like *Ctenopharyngodon idella, Hypophthalmichthys molitrix, Oncorhynchus mykiss, Parabramis pekinensis, Salmo letnica, Silurus glanis* and *Tinca tinca*. Others like *Carassius gibelio, Gambusia holbrooki, Lepomis gibbosus, Pseudorasbora parva* and *Rhodeus amarus* are present in the catches but in very low numbers.

6. Reference conditions

6.1 Reference conditions for rivers

Although past data for rivers in Prespa Lake watershed are very scarce, the *reference conditions* were more or less easy to determine due to: a) all rivers belong to the same river type; b) they have very short and rapid flows prior to inflow in Prespa Lake; c) their source waters belong to two different parks of nature where they are significantly protected from any important human activities; d) even with a limited number of samplings, water chemistry and biology were easily distinguished from the rest of the river water courses where human impact was much more severe.

Therefore, a relatively solid starting proposal for the reference conditions of rivers in Prespa Lake watershed would be very close to conditions found in water bodies Kranska 1 and Brajčinska 1, which are: natural water courses with good hydraulic contact with surroundings, rich riparian vegetation, clear water with very low conductivity (<100), slightly acidic, low in nutrients which are easily biodegradable, and with variable natural flora and fauna in and around the water courses (Tab. 9).

Reference conditions for the rivers in Prespa Lake watershed	
Parameter (units)	Value
Dissolved oxygen (mg·L⁻¹)	>9
Conductivity (μS·cm⁻¹)	<50
pH	6-7
NH₃-N (mg·L⁻¹)	<0.05
NOₓ-N (mg·L⁻¹)	<0.6
Total N (mg·L⁻)	<1.0
PO₄-P (mg·L⁻¹)	<0.020
Total P (mg·L⁻¹)	<0.030
Toxic heavy metals and priority substances (μg·L⁻¹)	<0.001
Dominant algae	Diatoms: *Meridion circulare, Meridion circulare var. constricta, Diatoma hyemalis, Diatoma mesodon, Eunotia spp., Staurosirella pinnata, Hannea arcus, Psammothidium daonense, Amphipleura pellucida, Decussata hexagona, Luticola nivalis, Diadesmis perpusila, Krsticiella ohridana, Pinnularia sudetica.* Red algae: *Lemanea fluviatilis.*
Dominant benthic invertebrates	*Heptagenia sulphurea, Baetis rhodani, Baetis alpinus, Baetis fuscatus, Baetis vernus, Potamophylax latipennis, Capnia vidua, Brachyptera risi, Nemoura cinerea, Austropotamobius torrentium, Astacus astacus*
DSFI index – invertebrates	≥7

Table 9. Reference conditions for rivers in Prespa lake watershed.

6.2 Reference conditions for Prespa Lake

Establishing the reference conditions for Prespa Lake (or any other lake) is a far more difficult task to perform. If the only reasonable and justified principle regarding every water ecosystem as a separate entity (the state-changed approach opposite to spatial state classification – Moss et al., 1997) is applied, than Prespa Lake cannot be compared for its reference parameters to any other lake (even with Lake Ohrid to which Prespa Lake is the major water source). This is even more important if the very turbulent and variable past of Prespa Lake is taken into account. Namely, the lake was formed by three rivers (underwater flows which are still detectable in the lake) which were constrained by karstic masses blocking their way to Ohrid Lake. Only then did the Prespa Lake ecosystem start to develop with a very variable surface area and volume in the past; there are numerous human constructions (buildings, roads) recorded at the lake's bottom today. All of these characteristics describe Prespa Lake as a very large water source body, intensively mixed by numerous sub-lacustrine sources of water and with very unstable water mass basically depending on climate, hydrologic regime and human activities. It is also a system in which there is a constant mixing of the water column, either by winds or powerful underwater currents and sources, which also means a constant supply of nutrients in the water column.

In a water body such as this, and combined with the classical lack of continual monitoring (especially regarding biology) data, establishing of reference conditions has been proven a formidable task. Nevertheless, we have succeeded in acquiring core samples dated from 10 ka BP and to briefly analyse for the first time the basic chemical (major cations, heavy metals, total N and P content) and biological (diatom assemblages) parameters in the core layers dated from 0.5, 1, 2, 5 and 10 ka BP respectively.

Regarding the concentrations of major cations and heavy metals obtained from the analyses of Prespa Lake core samples (Fig. 32), Prespa Lake is dominated by *aluminium* and *iron* throughout the analysed period. On the other hand, *calcium* has increased more than three times in the same period, as well as *sodium* in the last 500 years, while *potassium* by 30%. These are clear signs of human alterations of the natural conditions.

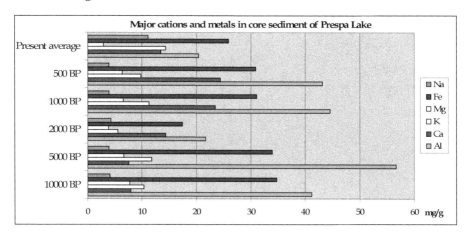

Fig. 32. Major cations and heavy metals in core samples of Prespa Lake.

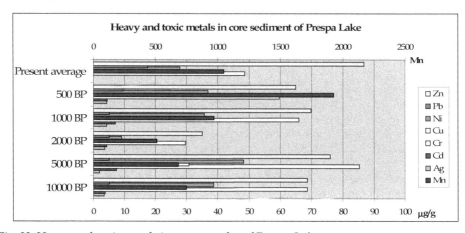

Fig. 33. Heavy and toxic metals in core samples of Prespa Lake.

Regarding heavy and toxic metals (Fig. 33) the greatest increase is recorded in concentrations of *zinc* and *manganese,* but *lead* is also showing a steady increase and a sudden peak in present times. These results are also corroborative of the obvious increase of human impact due to waste input in the lake's sediments in the past 500 years.

The results obtained for the total P content in the present day sediments of Prespa Lake (Fig. 34) are quite interesting. It can be concluded that the phosphorus in Prespa Lake has the crucial role in the overall eco-physiology of the system. It is not deposited at a regular pace and it is also not used in a predictive manner; a significant increase of phosphorous input has also been recorded during the summer months.

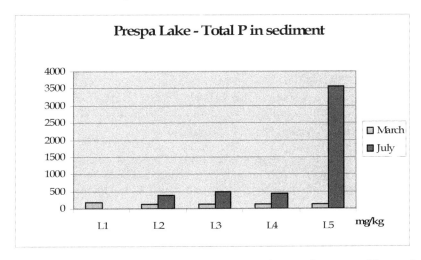

Fig. 34. Total P content measured in recent sediments at the sampling sites of Prespa Lake.

Compared to the results obtained from the analyses of the core samples (Fig. 35), the phosphorus in Prespa Lake reveals other important features. Firstly, it has been deposited in the recent sediments in significantly higher quantities (almost three times higher) than recorded in the core samples. Secondly, its predominance over nitrogen has been taking place in the last 500 years. Thirdly, Prespa Lake has never been a nitrogen limiting lake, since the values for total nitrogen are almost constant throughout the analysed period. Therefore, the principal nutrient that is driving the observed changes in the lake's plankton communities (cyanobacterial 'water blooms') is phosphorus. Observed occurrence of the cyanobacterial 'water blooms' at L5 sampling site (village Dolno Dupeni) and the results for the phosphorus deposition at the same area of the lake is more than just a coincidence which deserves much more attention in the future.

There are very few organisms, or their remains, that are well preserved in lacustrine sediments and can be easily retrieved for observations. Having siliceous cell walls, diatoms are probably the ultimate choice (Krstic et al., 2007) for both monitoring of the recent and paleo environments, since they quickly and constantly change their assemblages according to environmental conditions and their specific autecological preferences (Stoermer & Smoll, 1999).

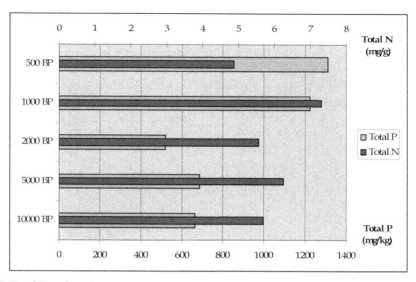

Fig. 35. Total P and total N Prespa Lake core sediments.

By analysing the diatom assemblages in different core layers of Prespa Lake presented on Figure 36, in order to reveal possible changes in dominant planktonic or benthic taxa and thus deduce the corresponding changes of environmental conditions forced by human activities, the following observations can be formulated:

- Diatom assemblages along the 10 ka core of Prespa Lake are surprisingly uniform. Only very slight changes in dominance of specific taxa can be observed; typically dominant throughout the core are *Cyclotella ocellata, Stephanodiscus rotula, Diploneis mauleri* and *Camplylodiscus noricus*.

- Diatom flora of Prespa Lake is very rich in taxa as recorded before (Levkov et al., 2006). But, the overall composition of taxa in the communities point to an ecosystem which is naturally rich in nutrients and enables development of diverse microflora which is reflecting the basic **mesotrophic state** (according to present state of knowledge regarding diatom nutrient preferences and autecology) of the environment at least up to 10.000 years BP.

- The only observed important occurrence of a diatom form that can be conclusive for a significant increase of nutrients in the ecosystem is the appearance of *Aulacoseira* spp. (especially *Aulacoseira granulata*) in the sediments approximately 1000 BP, and persisting in the communities to the present. This unique, but very subtle, change in diatom taxa dominance can be connected to the recorded high increase of phosphorus concentration in the Prespa Lake sediments presented on Fig. 35. For comparison, the *Aulacoseira* taxa determined in Prespa Lake can be found in co-dominance with various cyanobacterial taxa (which are usually regarded as potentially toxic) in the plankton of highly eutrophic lakes like Dojran Lake in Macedonia (Fig. 37 -Krstic et al., in prep). Since we cannot see the cells (or their remains) of other algae in the core layers, by deduction from the present knowledge we can conclude that Prespa Lake has become eutrophic, at least during the most productive periods, due to an increase of phosphorus and possibly other nutrients not yet analysed in the core samples. The presented time frame corroborates the strong possibility that human activities have played the crucial role in increasing of the trophic status of Prespa Lake.

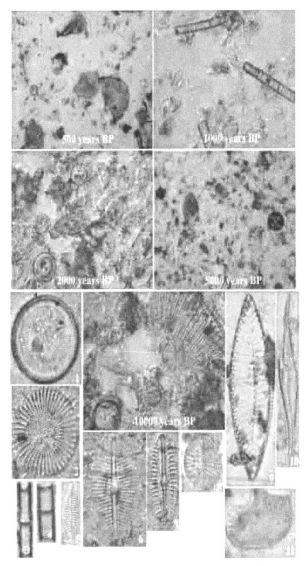

Fig. 36. Comparative presentation of diatom assemblages retrieved from 0.5-10 ka BP core samples of Prespa Lake and some of the most dominant and characteristic taxa in the investigated core samples: *1. Cyclotella ocellata, 2. Stephanodiscus rotula, 3. Aulacoseira granulata, 4. Aulacoseira ambigua, 5. Karayevia clevei var.balcanica* f.*rostrata, 6. Diploneis ostracodarum, 7. Diploneis mauleri, 8. Cavinula scutelloides, 9. Surirella bifrons, 10. Gyrosigma macedonicum, 11. Camplylodiscus noricus.*

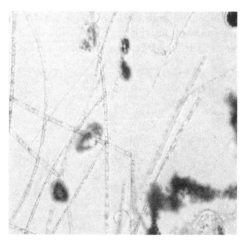

Fig. 37. Plankton sample from Lake Dojran (August 2010) dominated by *Aulacoseira granulata* and at least three *Microcystis* taxa; circular filaments belong to *Lynbya contorta*.

The final support for the overall conclusion that the Prespa Lake has completed the turnover to a highly eutrophic system came from the analyses of plankton communities during the summer months (Fig. 38). Only two cyanobacteria forms have produced a typical 'water bloom' from May until September, *Anabaena affinis* and *Anabaena contorta*, which have fully replaced the usual plankton dominance of diatoms belonging to genus *Cyclotella*. Consequently, ELISA tests for cyanotoxins (*microcystins*) in the lake's waters have revealed significant presence of these toxins in the summer months (Fig. 39).

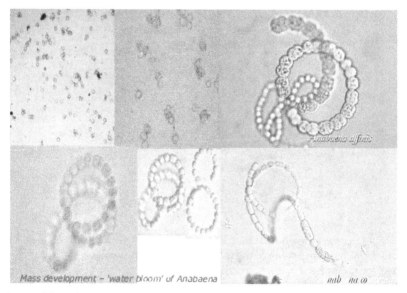

Fig. 38. 'Water bloom' caused by *Anabaena affinis* and *Anabaena contorta* in Prespa Lake waters.

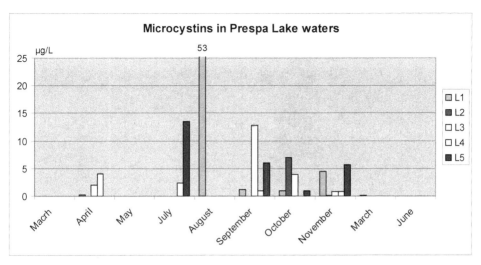

Fig. 39. Cyanotoxins-microcystins in Prespa Lake waters during the 12 month investigation period.

After all presented analyses and elaborations, the reference conditions for the Macro Prespa Lake ecosystem as one water body are presented in Table 10. Presented values for the most important parameters are targeted on the boundary between good and moderate water quality status for Prespa Lake which was surpassed at least a century ago. Having in mind the very high pressure of a variety of pollutants and human influences elaborated in this report, the targeted reference conditions may seem out of reach. But, if the ongoing situation continues, a total turnover of Prespa Lake towards highly eutrophic ecosystem should be expected in a very near future. In that case, the overall status of the Prespa-Ohrid-Crni Drim River system will be jeopardised and much more difficult to control, let alone brought to a good water quality status.

7. Discussion on the global significance of environmental alterations in lakes catchments

The people and their societies are integrated parts of the biosphere. They are dependent on its function and support, but at the same time they are shaping the biosphere globally with marked geological consequences (Steffen et al., 2011). This issue is much broader than the climate change *per se* (Folke et al., 2011). A key challenge for humanity in this new situation is to understand its role in the 'Earth System', start accounting for and governing natural capital and actively shape development in tune with the biosphere (Jansson et al., 1994; Rockström et al., 2009).

It is usually emphasised that during the last couple of generations we have witnessed an amazing expansion of human activities into a converging globalised society, enhancing the material standard of living for a large proportion of people on Earth (Rosling, 2010). This expansion has been quite pronounced since the 1950s, which predominantly benefitted the industrialised world, has pushed humanity into a new geological era, the *Anthropocene*, and generated the bulk of the global environmental changes with potential thresholds and tipping points, currently challenging the future wellbeing of the human population on Earth (Steffen et al., 2007).

Reference conditions for Prespa Lake	
Parameter (units)	Value
Dissolved oxygen ($mg \cdot L^{-1}$)	6-7 (surface); >4 (bottom)
Conductivity ($\mu S \cdot cm^{-1}$)	200-300
pH	7-8
NH_x-N ($mg \cdot L^{-1}$)	<0.05
NO_x-N ($mg \cdot L^{-1}$)	<1.0
Total N ($mg \cdot L^{-1}$)	<3.0
PO_4-P ($mg \cdot L^{-1}$)	<0.005
Total P ($mg \cdot L^{-1}$)	0.015-0.025
Chlorophyll a ($\mu g \cdot L^{-1}$)	<3.8
Secchi depth (m)	>5
Dominant algae	Diatoms, Chrysophytes, Green coccoid algae, Xanthophytes, Charophytes. No cyanobacteria or 'water blooms' by any algal group.
Dominant benthic invertebrates	Snails, Clamps, Dragon flies, Mayflies, Caddis flies, Leeches, Sponge, Amphipods, Decapods. No Chironomids or Tubificids indicators for eutrophic conditions
BQI index Diversity index H	>3 2.33-3.00

Table 10. Reference conditions for Prespa Lake.

The work presented in this chapter enabled a slightly different point of view which broadens the scope of human alterations of nature significantly beyond the last few generations. Namely, if a prolonged civilisation was present in a certain area, like Lake Prespa in our case, in which the geological conditions are fragile enough to critically influence the recipient waters (the lake in our case), than the anthropogenic history of impact is traceable. Unlike soils, water ecosystems can and will accumulate excess chemical compounds very rapidly, but will also rapidly release them in a given situation. This increasing of the 'total capacity' (Svirčev et al., 2010) of a water ecosystem inevitably alters the natural environmental balance and forces the accelerated eutrophication or the rapid aging of the ecosystem. In the case of Prespa Lake, the total deforestations (or the clear cuts –

Fig. 40) have been confirmed many times over in the recent past, but this is the first time they have been detected one millennium ago (Fig. 35). The change from the N driven system towards the P dominated one can only be attributed to intensive P leaching and washout of the karst based geology repeatedly deprived of significant vegetation cover. Human activities in the last 500 years have only added to the already initiated change of the Prespa Lake system.

Strategies for development that ignore the dynamics of the broader social–ecological system may push people into vulnerable situations and persistent traps, and undermine the capacity to sustain human wellbeing in the long-term. Science has responsibility to provide a better understanding of the challenges facing humanity and to explore pathways toward a sustainable world. Global and regional scale integrated assessments, inclusive, transparent, and founded on an understanding of social–ecological interactions play a central role in building momentum for 'Planetary Stewardship' (Folke et al., 2011).

Fig. 40. Clear forest cuts on Bigla Mountain, Prespa Lake watershed (contemporary practice by foresters).

8. Conclusions

The surface waters of Prespa Lake watershed are found to be under intensive human pressure. This pressure is expressed through various physical impacts like alterations of the water courses and water abstraction, chemical pollution originating from untreated wastewaters or agriculture, and deterioration of natural biodiversity by introduction of alien species or over fishing. The intensity and duration of the negative human impacts on Prespa

Lake's natural environment have resulted in severe deterioration of the water quality in almost all water bodies, except the elevated stretches of the rivers way beyond the immediate human activities, if medium range aero deposition is regarded as negligible.

Apart from the performed investigations and obtained results, one of the major contributions to the conducted analyses was via the method used to separate the natural background of nutrient emissions from the anthropogenic influences that need interventions. In the case of Prespa Lake, humans started to alter the environmental properties in its catchment more than 1000 years ago by intensive forest clearings that have resulted in accelerated phosphorus leaching. This process has become even more intensified in the past 100-150 years through untreated wastewaters inflow into the system and intensive agriculture. The final observed outcomes have been the full turnover of the dominant algae in the plankton towards cyanobacterial 'water blooms' during summer periods which have also proven to be toxic for microcystins.

In order to prevent further deterioration of the water quality in the watershed, substantial efforts have to be made and many water pollution prevention measures implemented. Even if these activities are fully implemented and operational, the timeframe for full recovery of the ecosystem may be prolonged, since the accumulated quantities of harmful substances are in the range of highly elevated levels. Nevertheless, if no measures are initiated and implemented in the area, the overall environmental quality in Prespa Lake watershed will be much more degraded in the near future. This is especially important for the Prespa Lake itself since it has already started to show clear signs of becoming eutrophic throughout the year with even more frequent and toxic cyanobacterial 'blooms'. If the turnover towards fully eutrophic system is completed, the activities to restore and improve its water quality in that situation will be much more difficult or even impossible, thus rendering Prespa Lake unsafe and unusable for future generations.

9. Acknowledgments

Many colleagues and contributors, as a part of the full team, have added their invaluable research data for achieving the final results. Among them, most important contributions for the purpose of this chapter have been made available by Prof.dr.Ivan Blinkov (GIS), Prof.dr.Ordan Cukaliev (agriculture), Prof.dr.Trajce Stafilov (heavy metals), Dr.Marina Talevska (macrophytes), Dr.Trajce Talevski (fish), Miss Mr. Radmila Bojkovska (priority substances), Mr. Valentina Slavevska (zoobenthos). The research has been conducted under the UNDP funded project "Development of Prespa Lake Watershed Management Plan", RF. 50/2009, Contract No. 31/2009, lead by Mr. Teodor Conevski. The author is fully obliged to all contributors.

10. References

Andersen, J. H.; Conley, D.J. & Heda, S. (2004). Implementation of the European water framework directive from the Basque country (northern Spain): a methodological approach. *Marine Pollution Bulletin*, 49: 283-290.

Birk, S. & Hering, D. (2006). Direct comparison of assessment methods using benthic macroinvertebrates: a contribution to the EU Water Framework Directive intercalibration exercise. *Hydrobiologia*, 566: 401-415.

Bode, R. W.; Novak, M.A. & Abele L.A. (1997). *Biological stream testing*. NYS Department of Environmental Protection; Division of Water; Bureau of Monitoring and Assessment; Stream monitoring unit; Albany., USA.

Boon, J.P. (1992). Essential elements in the case for river conservation. In: *River Conservation and Management*, Boon. P.J. Calow, P. and Petts, G.E. (Eds.), John Wiley and Sons Ltd., Chichester, pp. 11 - 33.

Braun-Blanquet, J. (1964). Pflanzensoziologie: Grundzüge der vegetationskunde, 3 ed. – Springer-Verlag, Wien. 865 pp.

BREF (2003). Reference Document on Best Available Techniques for Intensive Rearing of Poultry and Pigs.

CEN (2003). EN 13946 – Guidance standard for the routine sampling and pre-treatment of benthic diatoms.

Dinsmore, W. P.; Scrimgeour, G. J. & Prepas, E.E. (1999). Empirical relationships between profundal macroinvertebrate biomass and environmental variables in boreal lakes of Alberta, Canada. *Freshwater Biology*, 41: 91-100.

Downes, B.J.; Lake, P.S. & Schreiber, E.S.G. (1993). Spatial variation in the distribution of stream invertebrates: implications of patchiness for models of community organization. *Freshwater Biology*, 30: 119 - 132.

Folke, C., Jansson, A., Rockstrom, J. Olsson, P., Carpenter, S., Chapin, S., Cre´pin, A.S., Daily, G., Danell, K., Ebbesson, J., Elmqvist, T., Galaz, V., Moberg, F., Nilsson, M., Osterblom, H., Ostrom, E., Persson, A., Peterson, G., Polasky, S., Steffen, W., Walker, B. and Westley, F. (2011). Reconnecting to the Biosphere. AMBIO 40: 719-738. DOI 10.1007/s13280-011-0184-y

Grupce, Lj. (1997). Autochthonous and allochthonous quantities of phosphorus in Prespa Lake waters. *Symposium Proceedings*, 24-26.10.1997 Korcha, Albania, 68-78.

Hering, D.; Moog, O.; Sandin, L. & Verdonschot, P. (2004). Overview and application of the AQEM assessment system. *Hydrobiologia*, 516, 1-20.

ISO (2003). 7828 – Methods of biological sampling guidance for macroinvertebrates.

Jansson, A.M., M. Hammer, C. Folke, and R. Costanza (eds.) (1994). Investing in natural capital: The ecological economics approach to sustainability. Washington, DC: Island Press.

Krstic, S. (2007). Saprobiological and trophic models for Lake Prespa (saprographs) for use in similar regions and its application for evaluation of Ecological Quality Ratios (indicators). *EC-FP6 project "TRABOREMAINCO-CT-2004-509177, Deliverable 3.3.*, 98 pp.

Krstic, S.; Svircev, Z.; Levkov, Z. & Nakov T. (2007). Selecting appropriate bioindicator regarding the WFD guidelines for freshwaters - a Macedonian experience. *International Journal on Algae*, 9(1), 41-63.

Krstić, S. (2011). Estimation of the eutrophication level, consequences and remediation methods of some freshwater ecosystems in Republic of China – research on Lake Honghu and the east part of Lake Hubei. Final Project Report, *Ministry of Education and Science, Bilateral project R.China-R.Macedonia, Project number 03-1735/1, 2010-2011*, 50 pp.

Levkov, Z.; Krstic, S.; Metzeltin, D. & Nakov, T. (2006). Diatoms of Lakes Prespa and Ohrid (Macedonia). *Iconographia Diatomologica* 16: 603 pp.

Micevski, E. (2000). Geological and hydro-geological characteristics of the Ohrid - Prespa region. *Proceedings of the International Symposium "Sustainable Development of Prespa Region", Oteshevo 23-25.06.2000, Republic of Macedonia*, 10-17.

Moss, B.; Johnes, P. & Phyllips, G. (1997). New approaches to monitoring and classifying standing waters. In: *Freshwater quality: Defining the indefinable?* Boon, P.J, Howell, D.L.(eds), Scottish natural heritage, Edinburgh, 118-133.

Naumovski, T.; Novevska V.; Lokoska L. & Mitic V. (1997). Trophic state of Prespa Lake. *Symposium Proceedings*, 24-26.10.1997 Korcha, Albania, 132-137.

OECD (1982). Eutrophication of Waters, monitoring assessment and control. Paris, Organisation for Economic Cooperation and Development.

Rockström, J., W. Steffen, K. Noone, Persson, S.F. Chapin III, E.F. Lambin, T.M. Lenton, M. Scheffer (2009). A safe operating space for humanity. Nature 461: 472–475.

Rosling, H. 2010. Gapminder, for a fact-based worldview.
http://www.gapminder.org/.

Sandin, L. & Johnson, R. K. (2000). The statistical power of selected indicator metrics using macroinvertebrates for assessing acidification and eutrophication of running waters. *Hydrobiologia*, 422/423, 233-243.

Skriver, J.; Friberg, N. & Kirkegaard, J. (2000). Biological assessment of running waters in Denmark: Introduction of the Danish stream fauna index (DSFI). *Verh. Int. Ver. Limnology*, 27:1822–183.

Steffen, W., Persson, L. Deutsch, J. Zalasiewicz, M. Williams, K. Richardson, C. Crumley, P. Crutzen (2011). The Anthropocene: From global change to planetary stewardship. Ambio. doi:10.1007/s13280-011-0185-x.

Steneck, R.S., T.P. Hughes, J.E. Cinner, W.N. Adger, S.N. Arnold, S. Boudreau, K. Brown, Berkes, F. (2011). The gilded trap of Maine's lobster fishery: A cautionary tale. Conservation Biology. doi:10.1111/j.1523-1739.2011.01717x.

Stoermer, E.F. & Smol J.P. (eds.) (1999): *The Diatoms: Applications for the Environmental and Earth Sciences*, Cambridge University Press, Cambridge, 484 pp.

Stumm, W. & Morgan, J. (1981). Aquatic chemistry. New York: Wiley.

Svirčev Z, Marković S, Krstić S, Krstić K, Obreht I (2010) Ecoremediation (ERM) and Saprobiology – is there a link? Geophysical Research Abstracts Vol. 12, EGU2010-2052, 2010.

Rosenberg, D.M. & Resh, V.H. (1993). *Freshwater biomonitoring and benthic macroinvertebrates*. Chapman and Hall, New York, 488 pp.

WHO (1993). Assessment of Sources of Air, Water, and Land Pollution. A Guide to Rapid Source Inventory Techniques and their Use in Formulating Environmental Control Strategies.

WFD (2008). Directive 2008/105/EC – Priority substances. Official Journal of EU, 24 December 2008.

Van Dam, H. et al. (2005). *ECOSURV Project Report – BQE for Hungarian water bodies: Phytobenthos*, 54 pp.

5

Intervention of Human Activities on Geomorphological Evolution of Coastal Areas: Cases from Turkey

Cüneyt Baykal[1], Ayşen Ergin[1] and Işıkhan Güler[2]
[1]Middle East Technical University, Department of Civil Engineering,
Ocean Engineering Research Center,
[2]Yüksel Proje International Co. Inc.,
Turkey

1. Introduction

Coastal engineers, geomorphologists and scientists have been trying to understand and find the answer to one simple and a major question among a wide variety of challenging coastal problems for decades. Where will the shoreline be tomorrow? Or after a severe storm? Or next year? Or in a decade? In other words, how will the coasts of our earth evolve in time? And where do we stand as human being in this highly complex, yet fragile evolution of coasts?

Coastal areas are often highly scenic and offer plenty of natural resources. Over 50 percent of the world's population lives within 200 kilometers of coastline (Hinrichsen, 1994; Deichmann, 1996). According to the projections of future population, 75 percent of the world's population, or 6.3 billion people, could reside in coastal areas by 2025 (Hinrichsen, 1996). Increasing population and urbanization at coastal areas, unconscious exploitation of natural resources, incompetent management and education strategies and lack of control mechanisms increase the complexity of problems and severity of measures at coastal areas. One major coastal problem that almost every country with some kilometers of coastline faces and spends millions of dollars to solve the imbalance in coastal sediment budget at coastlines, resulting in severe erosion or accretion problems and loss of income from tourism and other coastal opportunities. Miami Beach, in Florida is a typical example for the above given problem. In the early 1970s, there was severe erosion at Miami Beach, where it was lined with seawalls, and compartmented by long steel groins. The numbers of visitors to Miami Beach and hotel occupancy rates were in decline. As a remedial measure, from 1976 to 1981, the Miami Beach Nourishment Project was undertaken, widening the beach by 100 m over a length of 16 km at a cost of $64 million USD. The project required in excess of 10 million cubic meters of sand obtained from offshore borrow areas by large hydraulic dredges. (Dean & Dalrymple, 2002). As a result, number of visitors at the beach increased from 8 million in 1978 to 21 million in 1983 (Houston, 1995a).

To control sediment budget at coastal areas, several types of measures are applied. These measures are categorized in two groups as hard measures (jetties, groins, detached breakwaters, seawalls, dykes etc.) and soft measures (beach nourishments, sediment traps, etc.). The hard measures involve the construction of solid structures within the intertidal zone to reduce wave energy and stop the sea from interacting with the hinterland. In contrast, soft measures generally avoid the solid structure approach, but use natural environments and sediments to reduce wave action (French, 2001; Dean, 2002). However, application of these measures usually fails or does not work as planned if applied without a clear understanding of the picture in the problematic coastal area and a well-designed and strictly implemented integrated coastal zone and river basin management structure.

This chapter gives brief information on the theoretical background of sources of sediment transport mechanisms and physical and numerical modeling attempts to understand these mechanisms. The governing parameters of these mechanisms and their temporal and spatial effects on the geomorphology of coastal areas are briefly discussed. Special emphasis is given to the numerical modeling of shoreline changes due to longshore sediment transport which becomes one of the governing mechanisms in the long term at coastlines where human induced activities put a pressure on the sediment supply resources e.g. dunes, cliffs and rivers. Two case studies of coastal erosion problems from Turkey are covered in this chapter, focusing on the numerical modeling of shoreline changes around coastal defense structures. The first case study is from northern coasts of Turkey, Bafra alluvial plain where the Kızılırmak River discharges into the Black Sea. It is one of the RAMSAR protection sites in Turkey. It has a rich biodiversity and is an important habitat for globally endangered bird species. The amount of sediment carried by the river has reduced drastically after the construction of flow regulation structures on the river, which led to serious coastal land erosion at the river mouth. The second case study, as another example to human induced type of coastal erosion problems, is from one of the touristic areas of Turkey, Side beaches in Antalya at southern coasts of Turkey. Alteration of the characteristics of the sea bottom topography due to touristic considerations led a serious erosion problem in front of the Perissia Hotel, in Side. For these two cases, the implemented remedial coastal defense measures, the shoreline changes after the implementation of these measures and applications of numerical modeling to predict these shoreline changes are covered within the case studies.

2. Sediment transport at coastal areas

Coastal areas are the environments with a very sensitive and dynamic balance governed by various complex physical processes. The shape of coastal landforms is a response of the materials that are available to these processes acting on them (Woodroffe, 2003). Among these processes, sediment transport plays an important role in continuous geomorphological evolution of coastal areas. This evolution mainly depends on the sources and sinks (losses) of the sediment budget of the coastal area. Major sources of sediment in coastal areas are weathering of cliffs by marine action (waves, tides and currents), wind and rain, freeze-thaw process and groundwater flow, sediment carried by rivers, dunes, wind-blown inland material, biogenic materials such as shells and coral fragments especially in some tropical areas and human induced sources like artificial beach nourishment, nearshore disposal of dredged soil and industrial waste tipping. The losses in the sediment budget can be listed as erosion by waves, tides and currents, submarine canyons, human extraction along rivers (sand mining) and at nearshore for commercial purposes or improve navigation, damming of rivers and streams, fishing by the use of explosives and trapping of sand on the upstream side of the coastal structures (CIRIA, 1996).

Sediment in coastal areas are transported in longshore and cross-shore directions by waves, by tidal and other steady or quasi steady currents like wind induced circulation, by winds from land (cliffs, dunes and inland), by rivers, small streams and gullies and also by human agencies. In temporal scale, effects of longshore sediment transport become more distinct in long term (years or decades) like recession of shoreline or accretion of sediments. As for the short term event like seasonal changes or significant storm events, cross-shore movement of sediments becomes the main actor in geomorphological changes in coastal areas such as bar formations in the surf zone (Kamphuis, 2000; Masselink & Hughes, 2003).

2.1 Numerical modeling of sediment transport

Prediction of morphological changes in coastal areas mostly depends on determination of long and short term effects of these transport mechanisms and changes in source-sink parameters of sediment budgets of coastal areas. Depending on the effective coastal processes, sediment budget knowledge and scope of the study, coastal sediment transport problems can be solved using physical modeling techniques (Hamilton & Ebersole, 2001; Wang et al., 2002; Gravens & Wang, 2007; Kökpınar et al., 2007), or with numerical morphological models expanding from one-dimensional (Hanson, 1987; Dabees & Kamphuis, 1998) to sophisticated 2D or 3D models (Roelvink et al., 2009), or with analytical and empirical approaches. Both empirical and analytical solutions give quick results to the scientists and engineers and are quite practical in early stages of understanding the coastal problems, computation of beach evolution and preliminary design studies, yet, for a detailed investigation of natural processes in coastal areas, physical or numerical models are commonly implemented. Compared to physical modeling, numerical models possess certain advantages in allowing the investigation of a wide range of parameters, adaptability to a variety of sites, economical operation, and the absence of scale effects (Hanson, 1987). These advantages sometimes may lead misuse of these models or misinterpretation of the results of these models due to lack of knowledge related to the actual site conditions. The uses of numerical models change for different spatial and temporal scales. For example, one-dimensional numerical shoreline change models are preferable for most of the cases due to their applicability in wide ranged temporal (from 1 year up to 20 years) and spatial scales (from 0.5 kilometer up to 10 kilometers), whereas more sophisticated 2D and 3D models are generally utilized for medium-term (up to 10 years) and short (a single storm up to 1 year) shoreline changes, respectively, due to their complexity and intensive computations (Hanson et al., 2003).

2.1.1 One-line theory

The fundamentals of shoreline change models were first established by Pelnard-Considere (1956), who set down the basic assumptions of the "one-line" theory, derived a mathematical model, and compared the solution of shoreline change at a groin with laboratory experiments. The basic assumption of "one-line" theory is the concept of "equilibrium beach profile", that cross-shore transport effects such as storm-induced erosion and cyclical movement of shoreline position associated with seasonal changes in wave climate are assumed to cancel over a long simulation period. In this concept, the change of shoreline position (in cross-shore direction) in time is due to longshore sediment transport and local sources or sinks along the coast (such as river discharges, beach nourishment or net cross-shore sand loss). The boundaries of "equilibrium beach profile" are the berm height (B) on land, maximum elevation of sediment transport during the uprush (Masselink & Hughes, 2003), and closure depth (D_C) at sea, which is limiting depth beyond which no significant longshore and cross-shore transports take place due to littoral transport processes (Mangor, 2004), (Fig. 1).

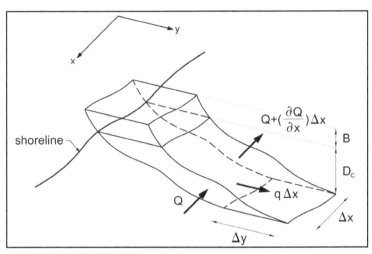

Fig. 1. Definition sketch for conservation of mass for a sandy beach system (Baykal, 2006)

Based on the above given assumption, "one-line" theory is expressed with the following differential equation which is also called as sand continuity equation,

$$\frac{\partial y}{\partial t} = -\frac{1}{(D_C + B)} \cdot \left(\frac{\partial Q_{t,lst}}{\partial x} + q \right) \tag{1}$$

where y is the shoreline position, x is the longshore coordinate, t is the time, $Q_{t,lst}$ is the total longshore sand transport rate, q represents local sand sources or sinks. Hallermeier (1978) relates the closure depth to $H_{s,12}$ and T_s which are the significant wave height exceeded 12 hours per year and corresponding significant period respectively (Eq.2).

$$D_C = 2.28 \cdot H_{s,12} - 68.5 \cdot \left(\frac{H_{s,12}^2}{gT_s^2} \right) \tag{2}$$

For the cases where the governing transport mechanism is assumed to be longshore sediment transport only, Hanson (1987) suggests to use the limiting depth of longshore sediment transport computed with the same equation (Eq.2) using significant breaking wave height ($H_{s,b}$) instead of $H_{s,12}$.

For the total longshore sediment transport rates including both bed and suspended sediment loads, the most widely used equations are Coastal Engineering Research Center of

The United States Army Corps of Engineers [thereafter CERC or USACE] (1984) and Kamphuis (1991) formulas. These formulas consider only wave generated currents and disregard other mechanisms like tidal or wind-induced currents. In CERC formula (Eq.3), the total volume of sand transported alongshore is related to the longshore component of wave power per unit length of beach,

$$Q_{t,lst} = \frac{K}{16 \cdot (\rho_s/\rho - 1) \cdot (1-p)} \cdot \sqrt{\frac{g}{\gamma_b}} \cdot H_{s,b}^{5/2} \cdot \sin(2\alpha_b) \tag{3}$$

where $Q_{t,lst}$ is the total volume of sediment moving alongshore per unit time (m³/sec), K is the dimensionless empirical proportionality coefficient given by Komar & Inman (1970) as $K_{SPM\ sig}$ = 0.39 based on field studies utilizing the significant wave height, ρ_s is sediment density taken as 2,650 kg/m³ for quartz-density sand, ρ is the water density (1,025 kg/m³ for 33 parts per thousand (ppt) salt water and 1,000 kg/m³ for fresh water), g is the gravitational acceleration (9.81 m/sec²) and p is the in-place sediment porosity that may be taken as 0.4. The breaker index (γ_b: ratio of breaking wave height to breaking depth) is taken as 0.78 for flat beaches (Weggel, 1972). $H_{s,b}$ and α_b are the significant breaking wave height and breaking wave angle respectively.

In the Kamphuis's (1991) formula (Eq.4), the effect of significant wave period (T_s), median particle size in surf zone (D_{50}) and the beach slope (m_b) at the depth of breaking, the slope over one or two wavelengths seaward of the breaker line are taken into consideration in addition to the significant breaking wave height and breaking wave angle.

$$Q_{t,lst} = 2.27 \cdot H_{s,b}^2 \cdot T_s^{3/2} \cdot m_b^{3/4} \cdot D_{50}^{-1/4} \cdot \sin^{3/5}(2\alpha_b) \tag{4}$$

Both CERC (1984) and Kamphuis (1991) formulas are utilized in shoreline evolution models extensively. Although these expressions have been verified in several researches for years, their predictive capabilities are limited mostly by site specific conditions (i.e. bathymetrical conditions, wave climate characteristics, sediment size and grading). Wang et al. (2002) state that Kamphuis (1991) formula produces more consistent predictions than the CERC formula for both spilling and plunging breaking wave conditions due to inclusion of wave period in the expression, which has significant influence on the breaker type. However, it seems most appropriate to use the CERC formula for high energy wave events and the Kamphuis (1991) formula for low-energy events (less than 1 m in wave height).

2.1.2 An example to one-line numerical modeling

For the purpose of understanding and predicting long term shoreline changes under the actions of wind waves, a numerical shoreline change model, CSIM (acronym for Numerical Model for Coastline-Structure Interaction), based on "one-line" theory is written in Ocean Engineering Research Center of Civil Engineering Department in Middle East Technical University (Şafak, 2006; Artagan, 2006; Baykal, 2006 and Esen 2007).

The numerical model is based on "one-line" theory with a built-in wave transformation module capable of simulating combined wave refraction and diffraction mechanisms in the vicinity of coastal structures such as groins and offshore breakwaters. It solves the sand continuity equation with both explicit and implicit finite difference schemes. The model can also simulate shoreline changes for some other coastal defense measure applications such as seawalls and artificial beach nourishment. The significant breaking wave heights in the surf zone are computed with formula given in CEM (2003);

$$H_{s,b} = (H_{s,0})^{4/5} \cdot [C_{g0} \cdot \cos(\alpha_0)]^{2/5} \cdot \left[\frac{g}{\gamma_b} - \frac{H_b \cdot g^2 \cdot \sin^2(\alpha_0)}{\gamma_b^2 \cdot C_0^2} \right]^{-1/5} \tag{5}$$

where $H_{s,b}$ is the significant breaking wave height, $H_{s,0}$ is the deep water significant wave height, α_0 is the deep water wave angle between wave crests and bottom contours, C_0 and C_{g0} are the deep water wave celerity and group celerity, respectively.

For combined effects of wave diffraction and refraction in the shadow zone of the coastal defense structures, Kamphuis's methodology (2000), which is based on Goda's diffraction diagrams (Goda et al., 1978), and refraction of regular waves are utilized. Combined effects of wave diffraction from multiple sources like consecutive offshore breakwaters are considered based on Vafaei's (1992) approach which is basically vectorial summation of wave rays coming from different sources of diffraction in proportion to the square of diffracted wave heights of these rays. More details and applications of the numerical model are given in Ergin et al. (2006) and Güler et al. (2008).

3. Case study-1: Coastal erosion problem at the Kızılırmak River mouth

In most of the developing countries, denser the population in coastal areas, the more vulnerable they become to severe environmental problems such as coastal erosion, exploitation and depletion of natural resources and extinction of endangered species. Wetlands at coastal areas are one of the most adversely affected areas due to their diverse floras and faunas. In Turkey, there are 13 sites designated as "Wetlands of International Importance" with a total surface area of 179,898 ha and 5 of these sites are located at coastal areas. One of these sites is Bafra alluvial plain (Kızılırmak Delta) where the Kızılırmak River discharges into the Black Sea. The site was designated as RAMSAR Area in 15.04.1998. It has a surface area of 21,700 ha including dunes, beaches, shallow lakes, seasonal marshes and wooded areas (URL-1). Numerous species of water birds, several of which are globally threatened, breed at this site. Over 92,000 water birds of various species winter at the site. In recent years, eutrophication, deforestation, illegal constructions and coastal erosion have become increasingly problematic in Kızılırmak coastal wetland (Kuleli et al., 2011). The location of Bafra alluvial plain is shown in Fig.2 and Fig.3.

The Kızılırmak River, which rises in the Eastern Anatolian Mountains, flows in a northwestern direction and discharges into the Black Sea by forming a conic alluvial delta (Fig. 2). It is the longest river in Turkey, with a length of 1,355 km, draining a basin of 74,515 km^2 (Kökpınar et al., 2007). The amount of sediment carried by the Kızılırmak River was 23.1 million tons/year till 1960's prior to any flow regulatory structures and decreased to 18 million tons/year following the construction of Hirfanlı Dam in 1960, and almost came to a cease with the total amount of 0.46 million tons/year after the constructions of Altınkaya Dam in 1988 and Derbent Dam in 1991 (Hay, 1994). This drastic decrease in the amount of sediment carried by the Kızılırmak River resulted in severe erosion with a maximum 1 km wide band of shoreline since 1988 according to the Regional Directorate of State Hydraulic Works and from local residents (Kökpınar et al., 2007).

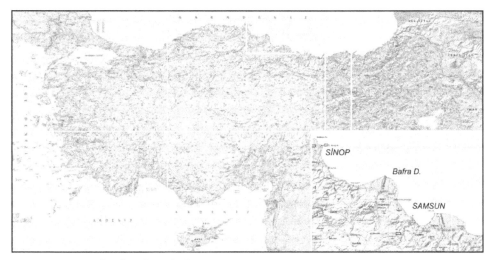

Fig. 2. Location of Bafra alluvial plain

Regarding the coastal erosion problem at Bafra alluvial plain, Kuleli et al. (2011) focused on the shoreline change rate analysis by automatic image analysis techniques using multi-temporal Landsat images and Digital Shoreline Analysis System (DSAS) along five Ramsar wetlands of Turkey. For Kızılırmak Delta, they have used three satellite images for the years 1989, 1999 and 2009 and found 16.1 m/year erosion rate for the Kızılırmak Delta.

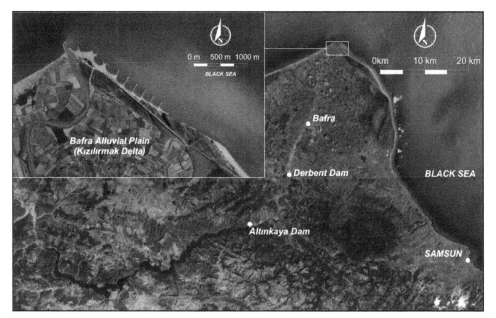

Fig. 3. Bafra alluvial plain and plan view of the existing shore protection system at the Kızılırmak River mouth (Google Earth, 2011)

The first remedial measure against this severe coastal erosion problem at the river mouth was held in 2000 by State Hydraulic Works (DSİ) based on the findings of the physical and mathematical model studies conducted at the Hydraulic Model Laboratory of DSİ in Ankara, Turkey (Kökpınar et al., 2007). It was composed of two Y-type and one I-type groins constructed at the eastern shoreline of the river mouth (Fig. 3). Ergin et al. (2006) studied the shoreline changes around these structures with the numerical shoreline change model, CSIM, using the shoreline measurements taken in 1999 (April) and 2003 (January) by DSİ. Despite utilizing T-groins instead of Y-groins and the numerical model's lack of capability of modeling tombolo formation, the model results was in good agreement quantitatively with the field measurements, especially at western sides (updrift) of second and third groins.

After the construction of first remedial system (two Y-type and one I-type groins), the shoreline retreat slowed down between the groins and trapping of sediment initiated. However, recession at the shoreline due to wave action continued to the east from the third groin (I-groin) as almost no sediment is carried by the Kızılırmak River. Later, two jetties were constructed at the west and east sides of river mouth between the years 2001-2004 to prevent seasonal closure of the river mouth. Between the years 2004-2005, the coastal defense system was extended with the construction five more I-type groins to prevent the collapse of drainage channel. Although, the drainage channel has been saved against wave action constructing the new series of five I-type groins, shoreline retreat at the east side of the defense system could not been prevented and continued to further east.

In this study, the performance of the second series of groins (5 I-type groins) and the shoreline changes around these groins between the years 2003-2007 are studied using the one-dimensional numerical shoreline change model, CSIM. The corresponding shoreline GPS measurements were provided by Regional Directorate of DSİ (Fig. 4).

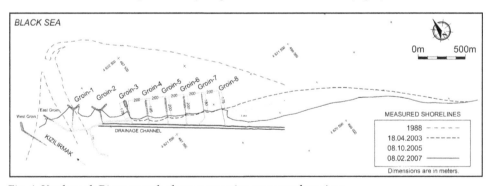

Fig. 4. Kızılırmak River mouth shore protection system plan view

3.1 Wave climate study: Kızılırmak River mouth

Long term geomorphological evolution of coastal areas under wave action results from series of short-term wave events occurring randomly. When there exist no time histories of these wave events or no continuous wave measurements, yet, wind measurements exist for such coastal areas, wave hindcasting studies are performed. For each wave direction, the wind velocities and effective fetch distances are used to hindcast the wave climate history of the site and a long-term wave statistics study is carried out to determine yearly or seasonal deep water wave characteristics.

There exist no continuous wave measurements for the shores close to Bafra alluvial plain. Therefore, a wave hindcasting study has been performed using hourly average wind data measured at 10 m above ground level by Sinop Meteorological Station between the years 2000-2009, obtained from General Directorate of Meteorological Affairs (DMİGM). The location of the river mouth is open to waves approaching from a wide directional sector from West to East-South-East. The effective fetch distances for the directions in this directional sector are determined from the navigation maps of Navigation, Hydrography and Oceanography Department of Turkish Naval Forces (SHODB). In the computation of effective fetch distances, for each direction, the effective generation area is considered as a sector from -22.5^0 to $+22.5^0$ totally covering an area of 45^0 with 7.5^0 intervals (USACE, 1984). The effective fetch directions and distances are shown in Fig. 5.

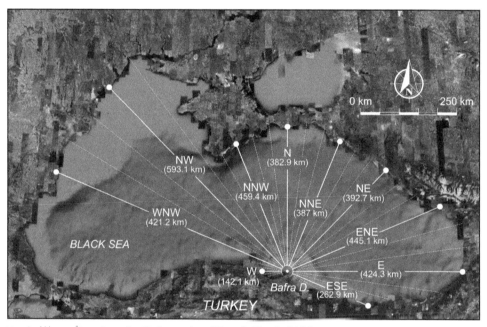

Fig. 5. Wave directions for Bafra region (Google Earth, 2011)

Using the effective fetch distances and the wind data obtained from Sinop Coastal Meteorological Station, deep water wave parameters (H_{s0} : deep water significant wave height, T_s : significant wave period) are obtained for the storms occurred during 10 years (2000-2009) by using the numerical model, W61, developed at Middle East Technical University, Department of Civil Engineering, Ocean Engineering Research Center (Ergin & Özhan, 1986; Ergin et al., 2008; Ergin et al., 2009).

The characteristic deep water wave steepness value (H_{s0}/L_0), the ratio of deep water significant wave height (H_{s0}) to the corresponding deep water wave length (L_0), for the project area is obtained as 0.040 from deep water significant wave heights and deep water wave lengths computed from corresponding significant wave periods (T_s) of each individual storm ($L_0 = gT_s^2/2\pi$ where g is gravitational acceleration in m/s^2).

To represent the actual wave conditions better and minimize the effect of wave data sequence in one-line numerical shoreline change modeling as discussed in Şafak (2006), a seasonal based wave data input is prepared. For this purpose, long term wave statistics is carried out for seasonal wave data. The hindcasted wave data is classified in 0.4m ranges for each season (winter season is assumed to include January, February and March, the other seasons followed accordingly) and the cumulative number of occurrences of each wave height class is plotted on to a semi-log graphical paper. The cumulative exceedance probability of deep water significant wave height, H_{s0}, is given as;

$$Q(> H_{s0}) = \exp[(H_{s0} - B)/A]$$

(6)

where $Q(>H_{s0})$ is the cumulative exceedance probability of a deep water significant wave height (H_{s0}). This equation indicates that if data points corresponding to H_{s0} and $Q(>H_{s0})$ are plotted on a semi-log graphical paper (H_{s0} on normal, and $Q(>H_{s0})$ on logarithmic scales), they should lie on a straight line with a slope of A and intercept of B when $Q(>H_{s0})$ is the horizontal axis.

Another major assumption in the preparation of wave data input is such that the effects of smaller but more frequent waves are considered to be more appropriate for a better representation of long term wave climate rather than higher waves with less frequency. By using the long-term statistics in seasonal time scale, the representative deep water significant heights ($H_{rs,0}$) of waves coming from every direction, their periods and seasonal frequencies in hours for each season are calculated using (Güler, 1997; Güler et al., 1998 and Şafak, 2006);

$$H_{rs,0} = \frac{\sum (P_i \cdot H_i)}{\sum P_i}$$

(7)

where H_i is the wave height and P_i is the occurrence probability of wave height H_i. Occurrence probability (P_i) of wave height (H_i) is computed by using the corresponding occurrence durations within the given range as follows;

$$P_i = Q(H_i - k) - Q(H_i + k)$$

(8)

where Q is the exceedance probability and k is an assigned range to compute occurrence probability. In Table 1, the seasonal wave data input for the model consisting of representative wave heights ($H_{rs,0}$), corresponding periods (T_s) and seasonal occurrence durations (Δt in hours) from all directions is presented.

As seen from Table 1, for all seasons the highest occurrences of waves within the range of 1 meter are from West-North-West and North-West.

3.2 Numerical modeling study: Kızılırmak River mouth

The shoreline changes between the second series of groins (5 I-type groins) constructed in addition to the first part (between years 2004-2005) starting from Groin-3 to Groin-8 with

200 meter spacing between adjacent groins (Fig. 4) are studied using the one-dimensional numerical shoreline change model, CSIM. The shoreline measurements for the years 2003 and 2007 are discretized at 10 m intervals. Starting from Groin-3, the apparent lengths of the groins are given as 175, 195, 200, 200, 180 and 175 meters respectively. From the sieve analyses of the sediment samples taken from the site, the median grain size diameter (D_{50}) is determined as 0.23 millimeters (Kökpınar et al., 2007). The berm height (B), the landward end of the active profile, is assumed as 2 meters. In the application of numerical modeling for the case in the Kızılırmak River mouth, it is assumed that no source or sink exists. The sequence of wave data input (Table 1) in the simulation starts with spring waves from W to ESE directions and continues with summer, autumn and winter waves.

Directions	WINTER			SPRING			SUMMER			AUTUMN		
	$H_{rs,0}$ (m)	T_s (sec)	Δt (hrs)	$H_{rs,0}$ (m)	T_s (sec)	Δt (hrs)	$H_{rs,0}$ (m)	T_s (sec)	Δt (hrs)	$H_{rs,0}$ (m)	T_s (sec)	Δt (hrs)
W	1.59	5.07	5.3	0.94	3.90	0.5	0.76	3.50	1.3	1.63	5.13	6.4
WNW	1.05	4.12	462.5	0.94	3.89	544.0	1.07	4.16	323.5	1.08	4.17	368.5
NW	1.08	4.17	391.4	0.86	3.72	149.2	1.18	4.36	119.9	1.05	4.11	221.2
NNW	1.03	4.09	100.1	0.72	3.41	113.1	0.81	3.62	172.3	1.19	4.39	124.0
N	0.91	3.83	66.7	0.58	3.06	16.3	0.70	3.36	41.4	1.13	4.27	13.5
NNE	0.83	3.66	14.4	0.70	3.35	17.0	0.63	3.20	22.4	0.84	3.67	28.6
NE	0.80	3.58	3.8	0.64	3.20	2.4	1.82	5.42	1.9	1.10	4.21	6.7
ENE	1.18	4.37	14.4	0.74	3.47	7.3	0.86	3.73	11.3	1.07	4.15	10.9
E	0.89	3.79	3.5	0.58	3.05	0.9	1.17	4.35	3.5	1.08	4.18	3.5
ESE	0.79	3.56	32.6	0.65	3.25	56.0	0.73	3.42	31.8	0.77	3.53	33.0

Table 1. Representative wave heights, corresponding periods and seasonal occurrence durations from all directions for each season

The sediment transport formulas are mostly based on various laboratory and field measurements. As the accuracy of these formulas may change from one site to another depending on the wave characteristics, grading of sediment, existing currents (tidal or wind induced) and sea bottom topography, direct use of these formulas may not reflect the actual sediment transport rates for the site studied. Therefore, calibration of these kinds of numerical models needs to be performed beforehand. For this study, the shoreline retreat at the downdrift of Groin-8 from 2003 till 2007 is used to calibrate the model.

The sea level variations of the Black Sea, where semidiurnal tides are dominant with a spring tidal range of 8-12 cm, has been largely controlled by a seasonal variation (inter-annual change) of 20–40 cm maximum in June (Alpar, 2009; Vigo et al., 2005; Bondar, 2007). The inter-decadal sea level variations have a period of 30 years and are in the order of 16 cm (Trifonova & Grudeva, 2002). The sea level variations produced by the atmospheric pressure changes or by sudden changes of wind direction could reach 10-20 cm with periods of 1-5 hours (Bondar, 2007). In the computations, the effects of these sea level variations are disregarded. The results of the simulations are given in Fig. 6.

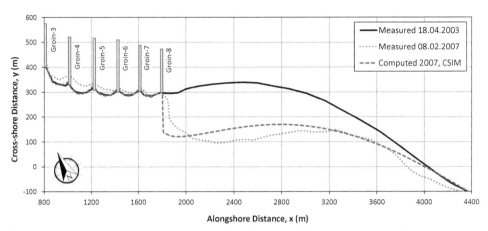

Fig. 6. Comparison of the site measurements with the numerical simulation (Groin series and downdrift of Groin-8)

3.3 Results and discussion: Kızılırmak River mouth

In the design of coastal defense structures such as shore perpendicular groins as in the case of the Kızılırmak River mouth, adverse effects of the defense structures at the adjacent shores have to be studied before the implementation of the project. In this case study, as an application of the numerical model, the shoreline changes both between coastal defense structures and adjacent shore are studied together.

Fig.6 shows that the shoreline changes between groins could only be simulated qualitatively; yet, the shoreline retreat after Groin-8 is in good agreement with the measurement quantitatively. For the groin system, the numerical model results reflect the shoreline changes qualitatively but not quantitatively. A reason for the inconsistency between the measured and computed shoreline positions for the year 2007 at the groin series might be due to the actual bypassing amount of sediment from the first 2 Y-type and 1 I-type groin system is not well defined. Also the small amount of the sediment carried by the river is neglected in the computations as no reliable knowledge or measurements regarding the amount of sediment carried by the river exist. As for the downstream shore adjacent to Groin-8, the agreement between the numerical model results and measurements are in good agreement both qualitatively and quantitatively.

4. Case study-2: Coastal erosion problem in Side, Perissia Hotel Beach

The second example for the adverse effects of anthropogenic activities on geomorphological evolution of coastal areas is from one of the touristic areas of Turkey; Perissia Hotel Beach, located in Side in Antalya at southern coasts of Turkey (Fig. 7). Side is one of the resort towns in Antalya, heavily utilized by touristic activities with numerous hotels. The closest rivers to Side are the Manavgat River (at 93 km in length, app. 12.5 km to Perissia Hotel Beach) and Köprü Çay stream, one of the finest rafting sites in Mediterranean region (at 14 km in length, app. 19 km to Perissia Hotel Beach). There are two dams on the Manavgat River; Oymapınar (1984) and Manavgat (1987) dams. The Manavgat River drains a topographic basin of 928 km² extending to the northern Taurus mountain range and with the addition of several closed basins (Mesozoic limestone aquifers), it has a total drainage area of 9,100 km², one of the major catchment basins in the Mediterranean coastal area (Yurtsever & Payne, 1986). The direction of the net longshore drift is from East to West.

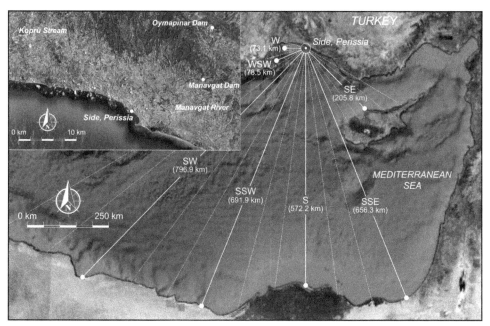

Fig. 7. Location of Side, Perissia Hotel beach and effective fetch distances for all directions (Google Earth, 2011)

The beach in front of Perissia Hotel is approximately 300 meters long and the width of the beach was approximately 50 meters before 1999. It is protected by naturally existing rocky formations at the east and west boundaries of the beach which extend 25 meters into the sea and a 120 meter long pier at the center of the beach, rocky formations lay underneath. The major protection for the beach against erosion due to wave attack was the rocky formations which were located at approximately 1-1.5 meter water depths and 60-100 meters offshore acting as submerged breakwaters and dissipating the energies of the approaching waves. However, the rocky formations in the western part of the beach were removed from the sea bottom in respond to customer needs in 1999. Altering the characteristics of the sea bottom topography, a serious erosion problem (retreating approximately 30 meters) was initiated at the beach mostly due to offshore movement of sands under wave action especially in winter season (Fig. 8). Remedial attempts utilizing artificial nourishment and construction of a groin of 25 meter long and 2 meter wide behind the pier could provide some amount of accretion at eastern part of the beach but did not work properly for the rest of the beach.

For the purpose of finding a consistent and effective remedial measure for the coastal erosion problem at Perissia Beach, site investigations and bathymetrical surveys were performed, wave climate of the region was determined and numerical modeling studies were carried out for alternative solutions by Güler et al. (2008). The wave climate of the region was determined carrying out a wave hindcasting study using the hourly average wind data measured by the nearest coastal meteorological station (Alanya Meteorological Station) for the years 1993-2004 which is obtained from DMİGM. The dominant wave directions were found to be South-South-East, South and South-South-West directions. Alternative solutions for the coastal erosion problem including hard and soft measures were discussed in detail in view of effectiveness.

Fig. 8. Perissia Hotel Beach at Side; shoreline retreat due to erosion over the years, red line shows the shoreline position approximately in August, 2006 (a: view of the western part of the beach from sea, b: view of the eastern part of the beach from sea; pictures from 1999; Güler et al., 2008)

The coastal protection system proposed was composed of several measures including both hard and soft measures, considering geomorphological features of the beach, sediment trapping capacity, possible harm to neighboring beaches and also scenic beauty. It was planned to be applied in three stages. The first stage is the construction of a 40 meters groin, 15 meters of which is at the land part, at the western border of the beach in addition to the existing groin at the mid-beach. This groin was planned to be a low crested structure, which will be at mean sea level at high tides and during storm events. It was actually planned to use the existing rocky formation at the western edge of the beach and strengthen it up to 25 meters seaward from the beach. The second stage includes the construction of a 40 meters submerged breakwater which is 30 meters away from pier and 60 meters offshore at the western part of the pier with a crest height 1 meter below still water level. The final stage was the nourishment of the western part of the beach between to groins (one existing and one constructed) at least 30 meters from the shoreline having a median grain size of around 1 mm (approximately 7000 m³ of sand). It was left optional to utilize gabion structures in front of the nourished area against wave action in cross-shore direction. The median grain size diameter (D_{50}) was found as 0.15 millimeters from the sieve analyses of the sediment samples taken from the site (Ergin et al., 2006). The summary of the proposed solution is shown in Fig. 9.

A recent site investigation was performed in January, 2010. Applied measures are investigated and the current shoreline position was measured. During the site investigation, it is found that the implemented remedial measures differ than the proposed coastal defense system presented above slightly in dimension and construction sequence:

- Western groin is extended to 15 meters towards the sea.
- Existing 25 meters long groin underneath the pier was removed for aesthetic reasons.
- Approximately 6000 m³ sand artificially added to the beach.
- The construction of the submerged offshore breakwater is delayed. The existing 25 meters long groin underneath the pier was removed.

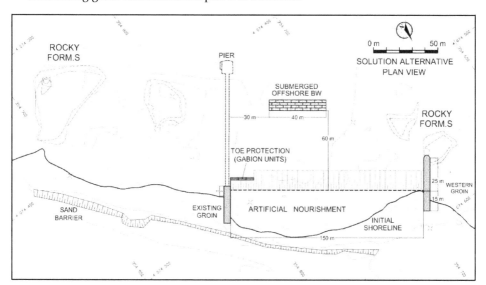

Fig. 9. Proposed remedial measures for the shoreline erosion at Perissia Hotel Beach at Side and nearshore bathymetry (18.11.2006)

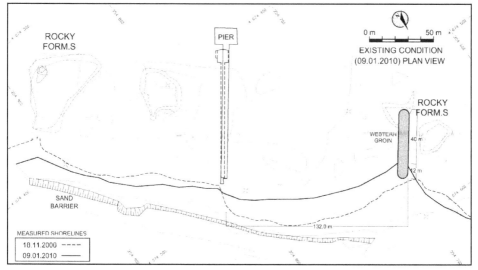

Fig. 10. Condition of Perissia Hotel Beach at Side at 09.01.2010, applied coastal defense structures and nearshore bathymetry (18.11.2006)

The implemented remedial structures and condition of the beach during the site investigation are shown in Fig. 10. In this study, the implemented measures are discussed and shoreline changes between the years 2006-2010 are studied with one-dimensional numerical shoreline change model, CSIM.

4.1 Wave climate study: Side, Perissia Hotel Beach

To obtain the wave climate in the last ten years at Side, a wave hindcasting study is performed using the using hourly average wind data measured at 10 m above ground level by Alanya Meteorological Station between the years 1993-2004, obtained from DMİGM. For each direction, the effective fetch distances are determined for each direction with an effective generation area as a sector from -22.5° to +22.5° totally covering an area of 45° with 7.5° intervals, from South-East to West using the navigation maps of SHODB. The effective fetch directions and the effective fetch distances are shown in Fig. 7.

Using the effective fetch distances and the wind data obtained from Alanya Coastal Meteorological Station, deep water wave parameters (H_{s0} : deep water significant wave height, T_s : significant wave period) are obtained for the storms occurred during 12 years (1993-2004) by using the numerical model, W61, developed at Middle East Technical University, Department of Civil Engineering, Ocean Engineering Research Center. The characteristic deep water wave steepness value (H_{s0}/L_0) for the project area is obtained as 0.042 from deep water significant wave heights and deep water wave lengths computed from corresponding significant wave periods of each individual storm.

The wave conditions in Mediterranean Sea are moderate compared to the Black Sea. There is not a very significant difference between the wave characteristics and their occurrences between the seasons. Therefore, for Side, an annual based wave data input sequence is prepared for the numerical model to carry out long term wave statistics and determination of representative deep water significant wave characteristics for each direction. The wave data input for the model consisting of representative wave heights, corresponding periods and annual frequencies for all directions are presented in Table 2.

Directions	$H_{rs,0}$ (m)	T_s (sec)	Δt (hrs)
W	0.77	3.42	11
WSW	0.84	3.59	119
SW	0.88	3.67	53
SSW	1.01	3.93	137
S	0.97	3.84	229
SSE	1.04	3.99	26
SE	0.88	3.67	4

Table 2. Representative wave conditions and annual occurrence durations for all directions (Güler et al., 2008)

4.2 Numerical modeling study: Side, Perissia Hotel Beach

For the numerical modeling of the applied coastal defense measure system, the shoreline at western part of the beach is assumed to be nourished till the end of the newly constructed groin at land side which makes approximately 5600 m² of nourishment area. It is assumed that the shoreline remained same till the construction of the western groin. The length of the western groin is 52 meters from the nourished shoreline of 2008. The rocky formations shown in Fig. 9 and 10 are modeled as submerged offshore breakwaters at various crest heights, widths and distances from shoreline. Submerged breakwaters or such rocky formations close to still water level force the waves to break and dissipate their energies and create sheltered wave fields behind the formations. The wave heights behind such structures are computed using diffraction terms and transmission coefficients ($C_t=H_t/H_i$, ratio of transmitted wave height to incoming wave height) in CSIM. Depending on geometrical properties of the rocky formations given in the bathymetry measured in 18.11.2006, the transmission coefficients are roughly computed using below given equation (CIRIA, CUR, CETMEF, 2007) where R_c is the crest height from still water level and H_s is the significant wave height in at the toe of the structure.

$$C_t = 0.46 - 0.3 R_c / H_s \quad \text{for } -1.13 < R_c / H_s < 1.2 \tag{9}$$

The berm height (B), the landward end of the active profile, is assumed as 1.5 meters according to the site investigations. The sequence of wave data input (Table 2) in the simulation starts from W to SE directions for each year. For the calibration of the numerical model, the shoreline retreat after the construction of the groin till January, 2010 has been used.

Alpar et al. (1995) gives a 0.21 meters spring range of semi-diurnal tidal variations together with a seasonal variation of 0.18 meters sea level variations for Antalya based on sea level measurements of Antalya station for the years 1935-1976. The effects of sea level variations are disregarded in the computations.

The computed position of shoreline using CSIM and measured shoreline positions for the dates 18.11.2006 and 09.01.2010 are given in Fig. 11.

4.3 Results and discussion: Side, Perissia Hotel Beach

As seen from Fig. 11, the shoreline change computed by the numerical model in the eastern part of the beach is in agreement with the measured shoreline both qualitatively and quantitatively. The erosion at the eastern part of the beach is due to the removal of the 25 meters long groin underneath the pier.

For the western part of the beach, the shoreline changes are in agreement qualitatively but not quantitatively. The accretion and erosion at the toe of the groin both on the eastern and western sides are in very good agreement with the measured shoreline in 2010 both qualitatively and quantitatively. The main reason for the quantitative disagreement between measured and computed results for the western part of the beach is due to disregarding the cross-shore sediment transport in the computations. In this area, which is bounded with the pier and the groin, the cross-shore sediment transport appears to be more effective and has to be included in the computations at the further stages of investigations. The excessive erosion in the western part of the beach could be reduced with the construction of the proposed submerged offshore breakwater (Fig. 9) which will reduce the cross-shore sediment transport.

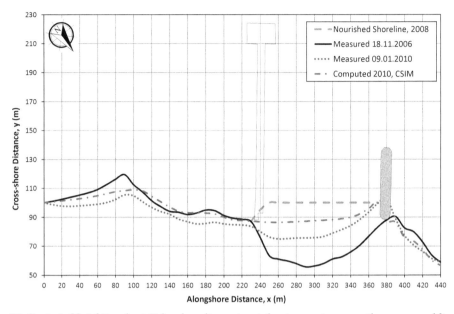

Fig. 11. Perissia Hotel Beach at Side; shoreline retreat due to erosion over the years, red line shows

Fig. 12. Two panoramic views of the Perissia Hotel Beach at 26.07.2006 and 09.01.2010 showing the before and after remedial measures

The coastal erosion problem faced with in Perissia Hotel Beach is a very important example of the erosion caused by alteration and harming of the natural balance in a coastal area, which results from damaging and removing the rocky formations, naturally acting as submerged breakwaters or reefs and thus, protecting the beach against wave action. Such an application has initiated and accelerated the erosion at Perissia Hotel Beach. In Fig. 12, the panoramic views of the beach at two different dates are given. The restoration of the western part of the shoreline is clearly seen in these pictures. However, the shoreline in eastern part started to retreat in return due to the removal of the groin underneath the pier. This retreat may be controlled by the implementation of the proposed remedial structures such as submerged offshore breakwater.

5. Conclusion

In this chapter, governing parameters of the geomorphological evolution of the coastal areas are briefly discussed in the view of sediment budget concept. Numerical modeling of the sediment transport mechanisms in coastal areas is discussed putting emphasis on one-dimensional numerical modeling which has been used as an effective tool by scientists and engineers for years to understand and predict shoreline changes in long term due to coastal erosion problems. In the case studies given above, the general approach for predicting shoreline changes at coastal areas, where human induced coastal erosion has become a chronic problem, is discussed. The Kızılırmak River mouth is a typical example of the effects of flow regulation structures on rivers and wetlands of alluvial plains fed by these rivers and resulting in severe coastal erosion problems in return. While silting of the reservoirs decreases the economic life of them, the coastal areas at downstream of these reservoirs hunger the silted material and degrades day by day. Perissia beach case is again another typical example of direct human intervention on the geomorphological and coastal-hydrodynamic processes at the beaches and resulting in erosion problems. These two cases stress the importance of the sustainable use of the resources in the long term for the benefit of mankind and of developing more sophisticated tools and methodologies to provide a better understanding and prediction capabilities of the adverse effects of anthropogenic activities at coastal areas.

6. Acknowledgements

The authors are thankful to General Directorate of State Hydraulic Works (DSİGM) and Bafra Plain Irrigation Project Directorate of DSİ for providing shoreline measurements of the Kızılırmak River mouth, to General Directorate of Meteorological Affairs (DMİGM) for providing wind data of Sinop and Alanya Meteorological Stations, to Navigation, Hydrography and Oceanography Department of Turkish Naval Forces (SHODB) for providing navigational maps used in computations. The authors also would like to thank to Prof.Dr. Ahmet Cevdet Yalçıner, Dr. Ilgar Şafak, Salih Serkan Artagan and Mustafa Esen for their efforts and supports in various stages of the above given studies.

7. References

Alpar, B., (2009). *Vulnerability of Turkish coasts to accelerated sea-level rise*, Geomorphology 107, pp.58–63

Alpar, B., Doğan, E. & Yüce, H., (1995), "On the long term (1935-1976) fluctuations of the low frequency and main tidal constituents and their stability in the Gulf of Antalya", Turkish Journal of Marine Sciences, Vol.1, pp.13-22

Artagan, S.S. (2006). A One-Line Numerical Model for Shoreline Evolution under the Interaction of Wind Waves and Offshore Breakwaters, M.S. Thesis, METU, Ankara, Turkey

Baykal, C. (2006). Numerical Modeling of Wave Diffraction in One-Dimensional Shoreline Change Model, M.S. Thesis, METU, Ankara, Turkey

Bondar, C. (2007). *The Black Sea level variations and the river-sea interactions*, GEO-ECO-MARINA 13/2007, Coastal Zone Processes and Management, Environmental Legislation

CIRIA (1996). *Beach Management Manual*, CIRIA Report 153, CIRIA, London

CIRIA, CUR, CETMEF. (2007). *The Rock Manual. The use of rock in hydraulic engineering (2nd edition)*. C683, CIRIA, London

Coastal Engineering Manual, "CEM" (2003). *U.S. Army Corps of Engineers, Coastal Engineering Research Center*, U.S.Government Printing Office Engineering, Yıldız Teknik Üniversitesi, İstanbul, Türkiye

Dean, R.G. (2002). *Beach Nourishment Theory and Practice.* Advanced Series on Ocean Engineering, Vol. 18, World Scientific Publishing, River Edge, N.J., 339 pp.

Dean, R.G. & Dalrymple, R.A. (2002). *Coastal Processes with Engineering Applications*, Cambridge University Press, Cambridge, UK (2002), 475 pp.

Deichmann U. (1996). *A Review of Spatial Population Database Design and Modelling.* National Center for Geographic Information and Analysis (NCGIA), University of California; Santa Barbara (UCSB), Santa Barbara, CA, USA

Ergin, A. & Özhan, E. (1986). *Wave Hindcasting Studies and Determination of Design Wave Characteristics for 15 Regions - Final Report*, Middle East Technical University, Department of Civil Engineering,, Feb., 1986, Ankara (in Turkish)

Ergin, A., Güler, I., Yalçıner, A.C., Baykal, C., Artagan A.A. & Safak, I. (2006). *A One-Line Numerical Model for Wind Wave Induced Shoreline Changes*, 7th International Congress on Advances in Civil Engineering, Istanbul, Turkey

Ergin., A., Yalçıner, A.C., Güler, I., Baykal, C., Şafak, I., Artagan, S.S., Esen, M., and Özyurt, G. (2007). *Side Perissia Perissia Hotel Beach Preservation and Control Project - Final Report*, Middle East Technical University, Department of Civil Engineering, Ocean Engineering Research Center, Ankara, Turkey (in Turkish)

Ergin, A., Yalciner, A.C., Guler, I., Baykal, C., Esen, M. & Karakus, H. (2008). *Fugla Beach Protection and Control Project - Final Report*, Middle East Technical University, Department of Civil Engineering, Ocean Engineering Research Center, Ankara, December 2008 (in Turkish)

Ergin, A., Baykal, C., Insel, I. & Esen, M. (2009). *Tsunami and Storm Surge Evaluation for Yenikapı and Üsküdar Stations - Final Report*, Middle East Technical University, Department of Civil Engineering, Ocean Engineering Research Center, Ankara, June, 2009

Esen, M. (2007). *An Implicit One-Line Numerical Model on Longshore Sediment Transport,* M.S. Thesis, METU, Ankara, Turkey

French, P.W. (2001). *Coastal Defences : Processes, Problems & Solutions.* Florence, KY, USA: Routledge, 2001.

Goda, Y., Takayama, T., & Suzuki, Y. (1978). *Diffraction Diagrams for Directional Random Waves*, Proc. 16th Int. Conf. on Coastal Engrg., ASCE, pp.628-650

Gravens, M.B., & Wang, P. (2007). *Data report: Laboratory testing of longshore sand transport by waves and currents; morphology change behind headland structures*, Technical Report, ERDC/CHL TR-07-8, Coastal and Hydraulics Laboratory, US Army Engineer Research and Development Center, Vicksburg, MS.

Güler, I. (1997). *Investigation on Protection of Manavgat River Mouth*, Yüksel Proje International Co. Inc., Research Project Report

Güler, I., Baykal, C., Ergin, A. (2008). *Shore Stabilization by Artificial Nourishment, A Case Study: A Coastal Erosion Problem in Side, Turkey*, Proc. of the 7th International Conference on Coastal and Port Engineering in Developing Countries (COPEDEC), Paper No: 90 ,Dubai, UAE

Hallermeier, R.J. (1978). *Uses for a Calculated Limit Depth to Beach Erosion*, Proc. 16th Int. Conf. on Coastal Engrg., ASCE, New York, pp.1493-1512

Hamilton, D.G., & Ebersole, B.A., (2001). *Establishing uniform longshore currents in large scale sediment transport facility*, Coastal Engineering 42 (3), 199–218.

Hanson, H. (1987). GENESIS: *A Generalized Shoreline Change Numerical Model for Engineering Use*, Ph.D. Thesis, University of Lund, Lund, Sweden

Hanson, H., Aarninkhof, S., Capobianco, M., Jimenez, J.A., Larson, M., Nicholls, R.J., Plant, N.G., Southgate, H.N., Steetzel, H.J., Stive, M.J.F. & de Vriend, H.J. (2003). *Modelling of Coastal Evolution on Yearly to Decadal Time Scales*, Journal of Coastal Research, Vol.19, No.4, pg.790-811

Hinrichsen, D. (1994). *Putting the bite on planet earth: rapid human population growth is devouring global natural resources.* National Wildlife Federation, International Wildlife Magazine's, September/October 1994 Issue

Hinrichsen, D. (1996). *Coasts in crisis. The earth's most biologically productive habitats are being smothered by development. Only coordinated international action can save them.* Issues in Science and Technology, 1996, 12, pp39–47.

Houston, J. R. (1995a). *Beach Nourishment.* Coastal Forum, Shore and Beach, Vol 64, No.1, pp21-24.

Kamphuis, J.W. (1991). *Alongshore Sediment Transport Rate*, Journal of Waterway, Port, Coastal and Ocean Engineering, ASCE, Volume 117, pp. 624-640.

Kamphuis, J.W. (2000). *Introduction to Coastal Engineering and Management*, World Scientific, Singapore-New Jersey-London-Canada

Komar, P.D. & Inman, D.L. (1970). *Longshore sand transport on beaches*, Journal of Geophysical Research 75 (30), pp.5914–5927

Kökpınar, M.A., Darama, Y., & Güler, I. (2007). *Physical and Numerical Modeling of Shoreline Evaluation of the Kızılırmak River Mouth, Turkey*, Journal of Coastal Research, Vol. 23, No. 2, pp. 445-456, ISSN 0749-0208

Kuleli, T., Güneroğlu, A., Karslı, F. & Dihkan, M. (2011). *Automatic detection of shoreline change on coastal Ramsar wetlands of Turkey*, Ocean Engineering Vol.38, pp.1141–1149

Mangor, K. (2004). *Shoreline Management Guidelines*, DHI Water and Environment, 294pp

Masselink, G. & Hughes, M. G. (2003). *Introduction to Coastal Processes and Geomorphology.* Edward Arnold, London, 354 pp.

Pelnard-Considere, R. (1956). *Essai de Theorie de l'Evolution des Forms de Rivage en Plage de Sable et de Galets,* 4th Journees de l'Hydraulique, Les Energies de la Mer, Question III, Rapport No. 1, pp.289-298.

Roelvink, D., Reniers, A., van Dongeren, A., de Vries, J. V., McCall, R. & Lescinski, J. (2009). *Modelling storm impacts on beaches, dunes and barrier islands,* Coastal Engineering 56(11-12), pp.1133-1152

Şafak, I. (2006). *Numerical Modeling of Longshore Sediment Transport,* M.S. Thesis, METU, Ankara, Turkey

Trifonova, E. & D. Grudeva. (2002). *Sea Level Surface Variations in Bourgas and Varna Bays.* Proc. of Second Int. Conf. on Oceanography of Eastern Mediterranean and Black Sea: Similarities and Differences of Two Interconnected Basins, TUBITAK Publishers, Ankara, Turkey, pp.151-155

USACE (1984). *Shore Protection Manual,* Department of the Army, U.S. Corps of Engineers, Washington, DC 20314.

Vafaei, A.R. (1992). *Mathematical Modeling of Shoreline Evolution in the Vicinity of Coastal Structures,* M.S. Thesis, METU, Ankara, Turkey

Vigo, I., Garcia, D., Chao, B.F. (2005). *Change of sea level trend in the Mediterranean and Black seas,* Journal of Marine Research 63, pp.1085–1100.

Wang, P., Ebersole, B.A., Smith, E.R., Johnson, B.D. (2002). *Temporal and spatial variations of surf-zone currents and suspended sediment concentration,* Coastal Engineering Vol.46, pp.175–211

Weggel, J. R. (1972). *Maximum Breaker Height,* Journal of the Waterways, Harbors and Coastal Engineering Division, Vol 98, No. WW4, pp.529-548.

Woodroffe, C.D. (2003). *Coasts, form, process and evolution.* Cambridge University Press, 623pp.

Yurtsever, Y., & Payne, B.R. (1986). *Time-variant linear compartment model approach to study flow dynamics of a karstic groundwater system by aid of environmental tritium (a case study of south-eastern karst area in Turkey),* in Gunay, G., & Johnson, A.I., eds., Karst water resources: IAHS Publication No. 161, pp.545–561.

Web references

URL-1: http://www.ramsar.org

6

Spatial and Time Balancing Act: Coastal Geomorphology in View of Integrated Coastal Zone Management (ICZM)

Gülizar Özyurt and Ayşen Ergin
Middle East Technical University,
Civil Engineering Department,
Ocean Engineering Research Centre,
Turkey

1. Introduction

Many coastal processes depend on the geomorphology of the coastal area. Although some areas are naturally prone to high risk, anthropogenic actions further alter their geomorphology, rapidly increasing the risk to coastal areas of disasters by disturbing the spatial and time balance of natural processes. It is a given fact that coastal zones are important social and profitable regions with high population densities. Thus, the management of these areas is critical but complex, calling for interdisciplinary approaches. Nevertheless, international and national agencies urge the application of integrated coastal zone management (ICZM) as the most efficient action for sustainable development in the face of diverse problems, such as climate change (IPCC, 2007). Restoring the balance of coastal landforms is one of the major aims of the ICZM process and the explanation of the geomorphological changes that occur on the coast is becoming increasingly important in order to manage coastal resources in a sustainable way (Woodroffe, 2002). While geomorphologic dynamics of coastal areas influence the character of society, the actions of society change the geomorphology at the same time. This is an iterative mechanism that has gained appreciation over the past century. Initially thought of as stand-alone impacts of human intervention on the shorelines, these impacts appeared as connected mechanisms through the dynamics of nature. In the end, they became a threat to human activity at many locations around the world. Additionally, time lag between human intervention and geomorphologic changes underlines the complexity of the management of coastal areas and the importance of scale as a concept for both ICZM and geomorphologic studies (Rotmans and Rothman, 2003).

The determination of spatial and temporal scales is necessary at the start of any research for the well-defined discussion of the results. Although both scales control the level of detail and accuracy of the research, the impact of scale selection on integrated assessments is a well-known but understudied challenge (Rotmans and Rothman, 2003) with the use of different scales having a significant impact on the results, such as leaving important interconnected elements of the system out of the research area (for example, not including tributaries or a river basin for a study of coastal erosion). Accordingly, local variability might be missed both in time and in space. The extent of the research area (whether it is a sand grain, a cliff, a coastal town or a region) is determined by a spatial scale using two axes: planform (also called 'long-shore') or profile ('cross-shore') (Woodroffe, 2002). A range of geomorphologic processes and human activities exist on both axes, influencing ICZM plans. The temporal scale is another multilayer factor that is important for the preparation of ICZM plans. Overall, geomorphologic scales exist across a wide range, from seconds to hundreds of years (Davies, 1993). On one side, human activities on coastal areas take a longer time to impact shorelines. First of all, most actions take a couple of years to carry out (for example, dams along river basins causing coastal erosion). Next, human use is expected to continue for 50 to 100 years (maybe more). This range of the temporal scale is also defined as the engineering timescale (French & Burningham, 2009). On the other hand, ICZM requires a long-term planning perspective, considering short-term benefits and solutions for urgent problems as well as integrating future risks, such as climate change. Although ICZM started as a form of shoreline management and flood risk planning, it evolved into a management concept, covering social, economic and ecological assets and including a diverse range of problems from natural disasters to man-made events - such as oil spillage - being generally accepted within coastal management literature (examples given in McFadden et al., 2007 such as Bower & Turner, 1997; Sorensen, 1997; European Commission, 1999; Kay & Alder, 1999; de Groot & Orford, 2000). Thus, different spatial and time scales exist within an ICZM plan (McFadden et al, 2007) and the geomorphology of coastal areas is one of the important parameters that define these scales.

2. Coastal geomorphology and ICZM

The coast is easily defined as one of the most diverse and dynamic environments found anywhere on earth. Many factors (geologic, physical, biological and anthropomorphic) are responsible for shaping the coast and carrying on its dynamic characteristics. Geological events created the sediment that formed the foundation of the modern coastal zone. Over time, various physical processes have acted on this pre-existing geology, eroding, shaping and modifying the landscape. These mechanisms are influenced by tectonics, climate, ecology and human actions. At the same time, many of these drivers can be affected by the evolution of the Earth's surface. Geomorphology is the science that focuses on the quantitative analysis of these drivers and these physical processes that shape the Earth's surface (Sanders & Clark, 2010). The nature of these processes depends strongly on the landscape or landform under investigation and the time scales and length scales of interest. The primary driving forces that cause change in most landforms include wind, waves, chemical dissolution, groundwater movement, surface water flow, glacial action, tectonics and volcanism.

Coastal geomorphology deals with the shaping of coastal features (landforms), the processes at work on them and the changes taking place (Bird, 2008). Understanding changes at the coast can require an examination of processes well outside the coastal zone, focusing on the interactions of coastal-zone features and hydrologic, meteorological and fluvial forces by means of sediment transport. Fluid dynamics produce sediment transport, causing geomorphic change to a continuous extent of temporal and spatial scales. On the other hand, the change of geomorphic features alters the boundary conditions for fluid dynamics. In turn, it produces further variations in sediment-transport patterns which again cause changes in geomorphic features. These processes happen over a wide range of spatial and temporal scales (Fig. 1). For example, interaction with the near shore profile changes the properties of waves (generated by offshore storms) when entering the coastal zone (Woodroffe, 2002). The resulting wave and flow characteristics control the cross-shore and long-shore variations. The characteristics of the bottom slope and the variations of waves and tides dominate the dynamics of sediment fluxes, causing changes such as erosion and accretion. On the other hand, small-scale processes control the turbulent dissipation of breaking waves, the bottom boundary layer and the bed form mechanisms that shape the local sediment flux (US Army Corps of Engineers, 2003).

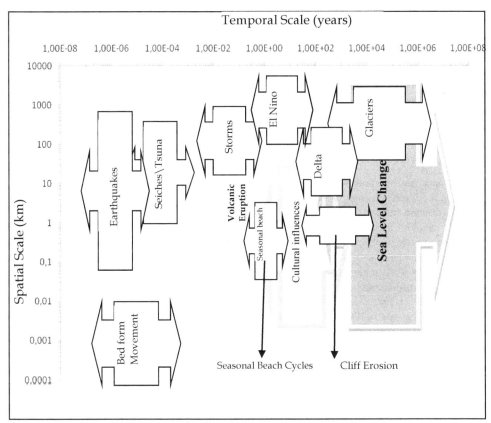

Fig. 1. Temporal and spatial scales of geomorphologic and coastal processes.

The rate of the response of geomorphic features to coastal processes depends on the scale, with larger features taking relatively longer to change. For example, under large waves significant changes in small-scale bed forms can occur within a single wave cycle, but changes in large-scale bed forms are established some time after the occurrence of the main driving force. Winds, waves, tides, storms and stream discharge are important driving forces in the coastal zone (Woodroffe, 2002).

In addition to the two main axes (long-shore and cross-shore), the coastal zone can be classified as micro-, meso- or macro-cell so as to define the spatial scale (Schwartz, 2005). Micro-cells include smaller geomorphic features such as ripples and small beach-face features which change over the period of a day or even hours. Meso-cells include geomorphic features such as beach profiles which change over a year or else months. Macro-cells extend for kilometres and include large coastal geomorphic features.

It is this macro-scale that presents one of the grand challenges of coastal geomorphology. Relating the prediction of morphodynamic behaviour at a meso-scale is of particular relevance for the understanding and management of coastal responses to environmental change. However, the study of coastal processes has traditionally been restricted to small and intermediate scales (Thornton et al. 2000 cited in French & Burningham, 2009), making it but one of several influences on the coastal zone (Fig. 2). For decadal-scale studies, coastal researchers use the 'one line' shoreline change model, which remains popular, especially for the prediction of the wider impact of engineering schemes. However, system linkages are rarely linear and numerical morphodynamic models are required to understand the quantitative response of the coastline to environmental forcing. Thus, meso-scale morphodynamic modelling is likely to be one of the most active research fronts in coastal geomorphology for the near future (French & Burningham, 2009). Unfortunately, at present the application of a single model cannot address most large-scale problems. Multiple models are typically required to achieve a reasonable qualitative and quantitative prediction of morphological changes (Hommes et al. 2007 cited in French & Burningham, 2009).

While many mechanisms such as human intervention, the climate and the sea level affect the coastal geomorphology in addition to coastal processes (Fig. 2), coastal geomorphology has a significant influence on those mechanisms as well. Depending on the characteristics of certain coastal landforms, most of the human activities that take place in coastal areas relate to tourism, agriculture and transportation (Woodroffe, 2002). Moreover, many structures are built on shorelines so as to ensure the sustainability of these activities. This cycle of affecting one another is what makes the design and implementation of ICZM complicated. Different uses of the same coastal resources might generate serious problems, especially under the threat of global forces such as climate change (IPCC, 2007). Also, the application of geomorphologic methods to predict the status of landforms usually covers small spatial scales and short term temporal changes as compared with scales of climate change and ICZM. ICZM practice requires longer time scales and larger spatial scales, and it relies on prediction of shoreline movements and landforms. There are various tools\models presented in the literature for different landforms and mechanisms, and it is the main challenge for ICZM practitioners to integrate and discuss the results of different models in order to come up with one plan for the coastal domain. That is why integrated assessment models have recently gained importance in coastal zone management research (McFadden et al, 2007). The threat of climate change and the impact related to sea level change has seen focus on longer term predictions as well. However, longer term predictions come with higher uncertainties. Thus, at the moment coastal policymakers need to select appropriate scales and tools so as to ensure the sustainable development of coastal areas, taking account of the different spatial and temporal scales of many mechanisms, each possessing different levels of uncertainties (IPCC, 2007). On the other hand, international agencies continue to call for integrated assessments of all disciplines – not just coastal processes - while much of the research focuses solely on one aspect of the question.

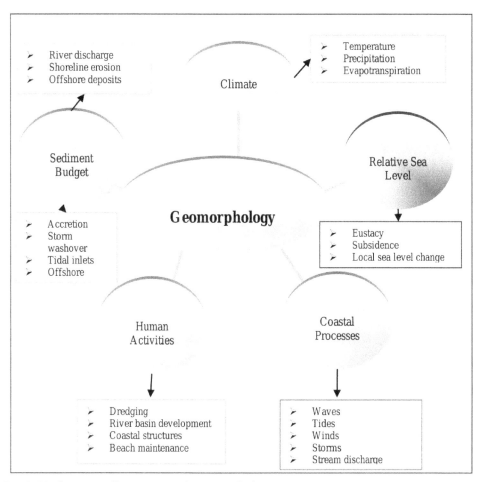

Fig. 2. Mechanisms affecting coastal geomorphology.

Vulnerability assessments are one of the tools used by coastal researchers to highlight the problem areas\sectors\processes that need management, both at the present and in the future. There are different vulnerability assessment methods which use only numerical modelling or highly qualitative procedures answering a range of questions from different perspectives (IPCC, 2007). Despite the use of computers, geographical information systems, remote sensing technologies, data management and support systems and developed methodologies, there remain many unknowns, uncertainties and challenges to overcome, both for the research community and for practitioners (IPCC, 2007; Klein and Nicholls, 1999). Most coastal areas lack continuous data collection or monitoring systems, which hinders the implementation of many of the available numerical models. For these regions, relying on the local experience of practitioners as well as historical events making up the concept of expert opinion is what the decision-making process generally corresponds to. Expert opinion - although a valuable tool - is subjective in nature when decision-making is considered. Keeping in mind that the wrong decision-making process could cause irreversible consequences for coastal areas, tools that could integrate expert opinion in an objective way with the available data and generic models and tools would mark an important step for the management of coastal areas where the monitoring and assessment of the implementation of developed plans could be scientifically verified. One such vulnerability assessment model was developed by Ozyurt (2007) and then upgraded in 2010 as a tool using fuzzy expert systems to integrate physical characteristics with human activities on both cross-shore and long-shore spatial extents for coastal areas (Ozyurt, 2010).

3. Coastal geomorphology and the Fuzzy Coastal Vulnerability Assessment Model (FCVAM)

The Fuzzy Coastal Vulnerability Assessment Model uses a total of 20 parameters (13 physical and 7 relating to human activity) to define such processes as coastal erosion, inundation, flooding due to storm surges and saltwater intrusion to freshwater resources (groundwater and rivers) along coastal areas. The Fuzzy Coastal Vulnerability Assessment Model (FCVAM) evaluates different areas through aggregated coastal vulnerability. At the same time, a region can be assessed for its vulnerability to different impacts. In addition, the vulnerability of governing physical\anthropogenic parameters is evaluated for individual impacts. The model uses an analytical hierarchy process to integrate stakeholder opinion with the decision-making process; the fuzzy expert system to combine expert opinion, the available data and generic coastal engineering knowledge and geographical information systems in order to present the results of the assessment of the vulnerability of coastal areas, integrating the impacts of sea level rise. The integration of important human activities - such as river basin management, land use and coastal engineering structures - with physical processes - such as waves, tides and sea level rise - on a process level is also achieved through the fuzzy coastal vulnerability model. Several spatial scales can be used for the implementation of the model and each spatial scale aims to be used for distinct purposes, directly related to coastal management practice. The implementation of the model to a coastal region would serve the purpose of a national vulnerability assessment determining "hotspots". Site-specific implementation, on the other hand, serves the purpose of developing a framework for adaptation planning.

The fuzzy coastal vulnerability model uses the main components of fuzzy expert systems, such as parameter membership functions, rule bases and fuzzy arithmetic, as well as clustering and data analysis tools such as the fuzzy c-means algorithm. The database used to develop the parameter membership functions is combination and integration of data covering European coastlines. Rule bases are developed by analysing the numerical models used to define coastal processes and the literature on climate change. The MATLAB Fuzzy Logic Toolbox is used to construct the fuzzy expert system, which is based on a modular structure enabling the future extension of the capability of the present model. Fuzzy arithmetic is also utilised in addition to a rule base expert system in order to determine the coastal vulnerability index. The details on the methodology of the fuzzy coastal vulnerability assessment model are presented in Ozyurt (2010). On the other hand, the focus of this chapter looks to discuss the role of geomorphology within the fuzzy coastal vulnerability model as well as the integration of spatial and temporal scales within the model in view of the different scales of geomorphology and ICZM through examples of application of the model.

Of the impacts assessed by the FCVAM model, coastal erosion is the one that both influences and is influenced by geomorphologic processes the most. The FCVAM model uses 6 parameters for physical characteristics and 5 parameters for anthropogenic activities in representing the mechanism of coastal erosion (Table 1).

	Physical Parameters	Human Influence Parameters
Coastal Erosion	1. Rate of Sea Level Rise	1. Reduction of Sediment Supply
	2. Geomorphology	2. River Flow Regulation
	3. Coastal Slope	3. Engineered Frontage
	4. Significant Wave Height	4. Natural Protection Degradation
	5. Sediment Budget	5. Coastal Protection Structures
	6. Tidal Range	

Table 1. Parameters of FCVAM model representing the coastal erosion mechanism (Ozyurt, 2010).

The physical parameters for erosion include and integrate the impact of climate change through sea level rise, geomorphology through landforms, and coastal processes through waves, the sediment budget and tidal range. Waves and tides are used to classify the shoreline as a high\low energy shoreline so as to determine whether or not coastal geomorphologic processes are governed by natural drivers. The sediment budget parameter - originating from the historical evolution of the shoreline - shows whether geomorphologic processes are dominant along the coastal area. The type of landforms points out the overall susceptibility of the shoreline to geomorphologic processes. All of the parameters mentioned are related to a long shore spatial scale.

The human influence parameters - also given in Table 1 - require assessments of their own which need to derive information from geomorphology studies as well. Due to activities outside the coastal zone, natural ecosystems (particularly within the catchments draining to the coast) have been fragmented and the downstream flow of water, sediment and nutrients has been disrupted (Nilsson et al. 2005). Land-use change - particularly deforestation - and hydrological modifications have had downstream impacts in addition to localised development on the coast. As stated by Jiongxhin (2004) erosion in the catchment has increased the river sediment load; for example, suspended loads in the Huanghe (Yellow) River have increased 2 to 10 times over the past 2000 years In contrast, damming and channelisation have significantly reduced the supply of sediments to the coast on other rivers through retention of sediment in dams. Indeed, this latter effect will likely dominate the 21st century. On the other hand, land use change along the shoreline (long shore scale) also changed the amount of sediment supply available to coastal processes. Excavation of the coastal zone, sand mining and urbanisation contribute to the change in the sediment budget because of human use of the coast. Human activities controlling the flow rate of rivers also have a significant impact on the supply of sediment to coastal areas.

Two parameters (the reduction of sediment supply and river flow regulation) are inserted into the FCVAM model in order to reflect these mechanical processes and the spatial scales of these mechanisms. The reduction of sediment supply is defined as the ratio of present sediment supply to the region to the natural state sediment supply in Ozyurt (2007). This parameter covers the sediment trapped in dams or reservoirs at the upstream of the river, the excavation of the coastal zone, mining and changes in land use. It defines the sediment particle itself and the abundance of it through different mechanisms, including the control structures on rivers that trap sediment. On the other hand, the river flow regulation parameter shows the degree of impact of any regulative structure on rivers at the down drift in terms of flow rate by using the methodology of Nilsson et al. (2005) relating to the flow regulation index (Ozyurt, 2007). This parameter focuses on the modification of the flow rate and the change in sediment movement along the river. As is well-documented in the literature, unregulated rivers carry the most sediment, partly because sediment is not trapped behind dams and partly because of the flushing of the river channel during floods or else high flow rates. While control structures enable stable flow rates, this change decreases the amount of sediment carried to the coastal area by generating favourable conditions for the settling of sediment particles along river channels (Ozyurt, 2010). The reduction of the sediment supply and river flow regulation parameters dictate the spatial scale (that is, if there is a river within the coastal zone then the basin is automatically included in the assessment as well as the associated processes influencing the geomorphology such as sediment transport). River basin management authorities can provide the necessary information for both parameters and this automatically secures the integration of other stakeholders - especially the ones along the river basin - as well.

Many structures, such as groins, seawalls, breakwaters and revetments, occupy coastal areas for different purposes, including the control of erosion and land loss. However, these structures themselves initiate undesirable impacts on the sedimentary processes of the region or neighbouring regions. Structures parallel to the coast - such as seawalls - cause the erosion of land in front of the structure and - in time - the whole of the land is lost if no other measure is put into practice. Although coastal protection structures may have a negative impact on adjacent shores, properly designed structures control the erosion and even initiate accretion within their region. Thus, coastal protection structures can decrease the vulnerability of the region when they work properly, achieving the intended results if adapted to sea level rise. If no adaptation measures are taken, these structures will lose their efficiency and, therefore, the vulnerability of the region will be increased.

Two parameters (engineered frontage and coastal protection structures) are included in the FCVAM so as to integrate the intervention of man-made structures on the coastal processes that alter coastal geomorphology. The engineered frontage parameter shows the percentage of the shoreline that the coastal structures occupies. This parameter includes all coastal structures (such as harbours, marinas, jetties and navigation channels) that do not have any purpose relating to the protection of the shoreline as well as coastal protection structures, which might cause adverse effects on adjacent shorelines. The coastal protection structures parameter, on the other hand, shows the percentage of shoreline that the coastal protection structures (such as groins and seawalls) occupy. As previously mentioned, coastal protection structures increase the resilience of the region if properly designed and adapted to sea level rise scenarios.

Another effect of human activities along coastal areas is the change in the ecosystem that increases the resilience of the coastal area, such as dunes and wetlands. However, these systems are under threat of urbanisation and other anthropogenic pressures. The natural protection degradation parameter shows the status of the ecosystem (such as dunes and marshes and wetlands) which provides protection for the coast. If the system is healthy, then the resilience of the area to the impact of sea level rise is high. For example, dunes act as both sediment supply sources against erosion as well as a barrier to inundation. If there is sand extraction from these dunes, although the area may be naturally resilient human activity significantly decreases this resilience. This change in the ecosystem also affects such the mechanisms as sediment transport which in turn impacts the geomorphologic processes.

In view of the aims of this study, the wave statistics and the spatial scale available for the region control the temporal scale which is used by the FCVAM for assessing coastal erosion vulnerability. For the coastal erosion process, the wave climate is one of the basic governing forces. Although single extreme events such as storms contribute to significant shoreline changes of a short duration, the coastal area always tries to establish equilibrium in the longer term. It is important to underline that storm-based coastal erosion might be more critical at some locations rather than long-term balance. For these locations, high resolution spatial scale and numerical modelling should be applied. This vulnerability might be what governs the natural hazard aspect of ICZM in the short temporal scale. However, if the time scale of sea level rise is considered, longer trends gain in importance. This is the reason for inserting another parameter (the sediment budget parameter) that represents historical and present shoreline movements so as to assess the vulnerability of coastal areas to coastal erosion.

The assessment of vulnerability due to storm surges is another module of the FCVAM (Table 2). Storm surges also have an impact on geomorphologic processes. However, the FCVAM model evaluates the flooding of coastal areas due to storm surges. Nevertheless, the time scale of the assessment - in terms of flooding - can be controlled through the storm surge height parameter by determining the return period (1, 10, 100, 1000 years).

Inundation, saltwater intrusion to groundwater resources and rivers are also included in the FCVAM, and geomorphology has influence on these processes as well (Table 2). Beach slope is the main parameter for the inundation mechanism, which is determined by the type of landform present at the coastal area. The properties of the soil layer and land use have an impact on the recharging of aquifers which, in turn, is included in the groundwater vulnerability assessment module. Also coastal geomorphologic processes significantly influence the geometry of the river mouth, which is represented by river depth in the river vulnerability assessment module. As can be seen, the parameters of the assessment are selected by analysing geomorphology studies for several time and spatial scales as well as other mechanisms. Thus, it would be appropriate to say that the upgrading of the assessment model is highly dependent on the advancement of geomorphology literature.

The next section will present examples of the application of the FCVAM model at different sites, focusing on the parameters directly related to geomorphology in addition to the concept of scale.

	Physical Parameters	Human Influence Parameters
Flooding due to Storm	1. Rate of Sea Level Rise	1. Engineered Frontage
Surges	2. Coastal Slope	2. Natural Protection Degradation
	3. Storm Surge Height	3. Coastal Protection Structures
	4. Tidal Range	
Inundation	1. Rate of Sea Level Rise	1. Natural Protection Degradation
	2. Coastal Slope	2. Coastal Protection Structures
	3. Tidal Range	
Salt Water Intrusion to	1. Rate of Sea Level Rise	1. Groundwater Consumption
Groundwater Resources	2. Proximity to Coast	2. Land Use Pattern
	3. Type of Aquifer	
	4. Hydraulic Conductivity	
	5. Depth to Groundwater	
	Level Above Sea	
Salt Water Intrusion to	1. Rate of Sea Level Rise	1. River Flow Regulation
Rivers/Estuaries	2. Tidal Range	2. Engineered Frontage
	3.Water Depth at	3. Land Use Pattern
	Downstream	
	4. Discharge	

Table 2. Parameters of the FCVAM model representing other mechanisms included in the model (Ozyurt, 2010).

4. FCVAM case studies: Viveiro, Spain, B'Buga, Malta and Silifke, Turkey

The FCVAM methodology can be applied at different spatial resolutions, depending on the objective of the assessment. The model can compare coastal areas from the NUTS3 level to local administrative unit levels with higher spatial resolution scales, since the model can analyse all of the combinations of model parameters having a range of input values. The application of the model is possible at a very coarse resolution such that aggregated vulnerability scores for different administrative units are used to evaluate national assessments and regional planning. At a local scale, it uses GIS-based high resolution data; the model analyses different impacts and the governing processes for a specific area. However, when the model is integrated with GIS-based information, the spatial scale of the assessment depends on the detail of the available information for a particular site rather than the extent of the research area. Both levels of assessment and how the different geomorphologic processes and landforms influence the model will be illustrated with case studies.

Three case study sites are chosen at the LAU2 (local administrative unit) level which corresponds to towns\municipalities at different countries. The model is applied to coastal strips within administrative borders using information from the database built by Ozyurt (2010) and the published research on these locations. The studies are also used to verify and validate applicability of the FCVAM model to different coastal areas and how the results can be applied to regional planning. The sites chosen are the Viveiro in Lugo province of Spain (Northern Spain), the B'Buga municipality of Malta (Southeast of Malta) and the town of Silifke, Turkey (south Turkey). These sites represent different landform types as well as different levels of human intervention on the geomorphologic processes from the perspective of ICZM planning and coastal vulnerability to sea level rise.

4.1 Database

The database build by Ozyurt (2010) covers most of the European coastlines from the Baltic Sea to the Atlantic Ocean and the Mediterranean Sea and include information on 79 major river basins and the aquifers of nine EU countries. This variety of coastal properties ensured the compilation of a thorough dataset enabling the application of the model to different coastal areas around the world. The database includes information on all the parameters of the FVCAM model presented in Table 1 and Table 2. Some of the databases used by Ozyurt (2010) form a part of other databases, which are either publicly or commercially available. However, all of the data collected and used by Ozyurt from these databases is available free for research. Some of the studies which were used to develop the database of Ozyurt (2010) were the EUROSION project, the DIVA project, the Digital Dataset of the European Groundwater Resources, the RivDIS dataset, the Waterbase dataset, the WWDII dataset and several national datasets. Details on the representation of different parameters within the developed database and the processes used to develop the GIS-based database are explained in Ozyurt (2010) in detail. The spatial resolution of the compiled dataset is not homogenous throughout European coasts (for some parameters, only information at a coarser resolution is available).

4.2 Case study areas: Viveiro, Spain

Viveiro (also known as Vivero) is a town and municipality in the province of Lugo, in the north-western Galician autonomous community of Spain. It has a residential population of over 16,000 (2010 figures), which triples in the summer months with visitors to the coastal region. Viveiro Ria is open to the north and is separated from the Barqueiro Ria by Coelleira Island (Fig. 3) on the lee side of Cape Estaca de Bares. The Landro River has developed an estuary in the inner part of the inlet. Its mouth complex used to present large sand spit, growing eastward from Covas headland. The Viveiro Ria is significantly affected by human occupation. The area includes the important fishing seaport of Celeiro in addition to extensive urbanisation and industrialisation. This occupation affects a large part of the marshland - most of which has been reclaimed - resulting in its current degraded state. Additionally, the highly modified Covas beach has developed between Punta Anchousa to the west and a dike to the east. The construction of the dike and the occupation of the dunes have resulted in a considerable erosion of sand. This loss has been compensated for with artificial regeneration of the beach. Research by Lorenzo et al. (2010) states that "the most significant changes on the beach and spit system have been (1) complete occupation of the dune area of the bar, (2) infilling of the former channel and of the Celeiro inlet, (3) construction of the seawalls of the Celeiro port, and (4) channelling of the Landro River outlet." On the other hand, the coastal strip of the administrative unit is dominated by rocky and medium cliffs, where the cliffs on the eastern side of the bay show signs of erosion. Small pocket beaches also exist along this indented coastal strip of the administrative borders of Viveiro.

4.3 Case study areas: B'Buga, Malta

The Maltese Islands have a collective shoreline of about 190 km and a surface area of 316 squares kilometres. Rough estimates indicate that only approximately 1.2% of the total land surface is 1m or less above sea level. In fact, the islands' coastline is characterised by cliffs, clay slopes and boulder rocks (Fig. 4). 50% of Malta's coasts and 74% of Gozo's coastline have been defined as inaccessible, mainly due to physical features (Malta Structure Plan, 1990 as cited in Axiak, 1992). This leads to heavy pressures being exerted on the remaining lowlands for touristic, industrial and urban purposes. Sandy beaches constitute only 2% of the coastline. Nonetheless, these very restricted localities and the rest of the coastal lowlands support a number of unique and important habitats, such as saline marsh lands, sand dunes and gentle rocky slopes. Coastal erosion is one of the human-induced pressures, including urban settlement and coastal development, land-based pollution and quarrying activities on the coastal lowland (Axiak, 1992).

Fig. 3. Google Earth image of Viveiro, Spain

Birzebbuga is a fishing port and small resort on the western side of Marsaxlokk Bay (Fig. 4). Its population increased because of the workers employed at the nearby Malta Freeport and container terminal. Popular among Maltese holiday-makers, the location is known for its important archaeological sites and a sandy beach. However, the coastal strip is mainly dominated by rocky cliffs and the heavy coastal structures of the Malta Freeport and container terminal.

The water supply in Malta is of two types. First class water is treated to potable standards and used for tap water. Second class water is non-potable and used mostly in agriculture and by some industries. First class water is derived from groundwater and five Reverse Osmosis (RO) desalination plants. These RO plants currently account for most of the water production and were introduced because of the high salinity levels in the mean sea level aquifer (Birdi, 1997). Groundwater is extracted from two main water tables. Almost all of the groundwater supplies (95.1%) are extracted from the mean sea level aquifer, a freshwater lens resting on denser seawater, and a small amount is obtained from the perched aquifer (4.9%), resting on a Blue Clay aquiclude in the west of Malta. Large quantities of groundwater are also extracted by farmers, industries and the private owners of boreholes. Unsustainable extraction policies, particularly during the 1970's, have meant that the water table is presently 40% over-exploited (Birdi, 1997).

Fig. 4. Google Earth image of B'Buga. Malta

4.4 Case study areas: Silifke, Turkey

Most of the important economic activities take place on or near the coastal areas and especially on deltas. While low elevations increase physical vulnerability, the high level of socio-economic activity exacerbates the vulnerability of these areas. One of these regions is the Goksu delta, which is located on the south of Silifke, Mersin, where the Goksu river - with a 10000-km2 catchment area - reaches the Mediterranean Sea. The coastal area of Silifke is dominated by the formation of the Goksu delta. The delta - surrounded by the Taurus Mountains to the north and northeast - is split into two by the Goksu river. There are two shallow lakes: Paradeniz and Akgöl, to the east and west respectively (Fig. 5).

The Goksu delta is known for the important biodiversity of its flora and fauna, which led to the Specially Protected Area status that it claimed in 1991. In 1994, the wetlands of the delta were included in the RAMSAR list (List of Wetlands of International Importance). The richness of the fauna of the Goksu delta is influenced by its geographical location as well as its ecology. The presence of the major sea turtles (Caretta caretta) nesting beaches on the Mediterranean and its importance in terms of ornithology make the delta one of the most diverse and valuable ecosystems in the region. In addition to its important ecological properties, the delta has become an important agricultural area, leading to rapid socioeconomic development since the implementation of irrigation network in 1968. The use of the river as a freshwater resource as well as the aquifers of the region has increased substantially after the 1980s when rice production started to dominate the agricultural landscape. High demand for freshwater - especially for groundwater - has led to sharp decreases in the water tables and the intrusion of sea water into some of the coastal aquifers. Across the region there are 5 municipalities and 7 villages, and the population is increasing with a rate above the average rate of the country. All of the economical, physical and ecological properties of the Göksu delta demonstrate the importance of this low-lying land, which has an average elevation of 2m above sea level (Ozyurt & Ergin, 2009).

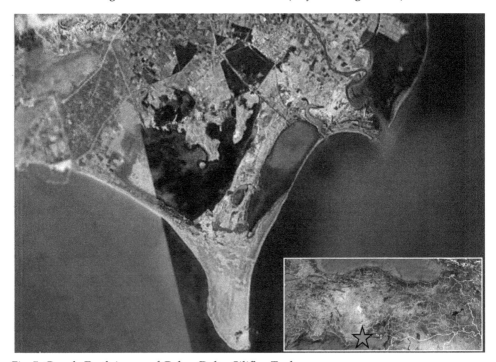

Fig. 5. Google Earth image of Goksu Delta, Silifke, Turkey

5. Results of the FCVAM Analysis

The FCVAM analysis is performed for the three case study areas by using the inputs collected from database of Ozyurt (2010) and published research papers on the specific sites

(Axiak, 1992; Birdi, 1997; Lorenzo et al, 2010; Ozyurt and Ergin, 2009). Table 3 presents the input values for the three study areas. Some of the data presented in Table 3 needs pre-processing steps. These pre-processing steps are explained in detail in Ozyurt (2010).

Physical Inference Parameters	Viveiro	B'Buga	Goksu
Rate of Sea Level Rise (mm/year)	2.5	1	2
Geomorphology	Medium cliffs	Rocky cliffs	Delta
Beach Slope (%)	2.6	3	1
Significant Wave Height (m)	2.96	1.04	3
Sediment Budget (%)	-17	-3	-50
Storm Surge Height (m)	6	3	4
Tidal Range (m)	2.75	0.1	0.3
Proximity to Coast of aquifer (km)	NA	0.4	0.4
Type of Aquifer	NA	Unconfined	Unconfined
Hydraulic Conductivity of aquifer	NA	0.001	0.000016
Depth to Groundwater Level (m)	NA	0	2
River Discharge (m3/s)	7.5	NA	90
River Water Depth (m)	1	NA	1
Human Inference Parameters			
Reduction of Sediment Supply (%)	30	0	60
River Flow Regulation	Not affected	NA	Moderately affected
Engineered Frontage(%)	27	10	5
Groundwater Stress(%)	NA	140	80
Land Use	Unclassified	Settlement\Industry	Agriculture
Natural Protection Degradation (%)	17	NA	60
Coastal Protection Structures (%)	9.3	10	3

Table 3. Input data used for the FCVAM Analysis.

The results of 3 case studies are given in Table 4. These values should be discussed using a scale of 1 to 5 which the FCVAM model also translates into levels of vulnerability, such as very low, low, moderate, high and very high (Ozyurt, 2010).

The coastal vulnerability assessment of the three different case study sites shows that Viveiro and B'Buga show moderate vulnerability to the impact of sea level rise and that the Goksu Delta (Silifke) shows a high vulnerability according to the aggregated vulnerability scores. The results are compatible with the literature on the impact of sea level rise on coastal landforms presented in the IPCC Assessment Reports (2007). The IPCC Assessment Reports (2007) generalise the vulnerability of coastal areas to sea level rise in terms of coastal landforms; cliff and indented coastlines inherit moderate vulnerability while deltas and low lying lands show high vulnerability, particularly to the impact of sea level rise.

	Regions					
Impacts	Viveiro		B'Buga		Goksu	
Coastal Erosion	3.00	Moderate	2.14	Low	4.00	High
Inundation	3.20	Moderate	3.00	Moderate	4.05	High
Storm Surge	4.00	High	3.00	Moderate	4.00	High
Groundwater	NA	NA	4.01	High	3.83	High
River	2.00	Low	NA	NA	2.88	Moderate
VULNERABILITY INDEX	3.05	Moderate	3.09	Moderate	3.71	High

Table 4. Results of the fuzzy vulnerability assessment model for the three case study sites.

The first case study area is the Viveiro region, where medium cliffs and indented coastal strips with pocket beaches dominate the geomorphology. Since the assessment is performed at the LAU2 level, the information on the geomorphology of the region has been determined using the most dominant characteristic, namely medium cliffs and indented coasts. This geomorphology indicates the low vulnerability of the coastal area as a region. Moreover, the aggregated vulnerability score shows that the region is moderately vulnerable to the impact of sea level rise. Moderate vulnerability is assigned to coastal erosion and inundation for the region where the high resilience of the geomorphologic characteristics of the coastal strip to withstand the high energy of such driving forces as tides and waves. The high vulnerability of the region to storm surges (documented by measurements (Lorenzo et al. 2007)) highlights the dominance of the driving forces of geomorphologic processes. Impact vulnerabilities are governed by the physical characteristics of the region and for processes along shorelines, and human intervention adds to the vulnerability to a moderate level. The assessment - using aggregated information for a large spatial area - is not able to highlight the variance of vulnerability across the region since the possible higher vulnerability of the pocket beaches were not reflected in the results. However, studies on the dynamics on the main beach (Cove beach) of the area shows that the beach is mostly stable because of annual artificial nourishment. Nonetheless, the moderate vulnerability score for coastal erosion is sufficient to explain the vulnerability of the coastal zone of the region, although the application of low resolution data misses the variance in vulnerability at higher spatial resolution. Additionally, due to the variability of the vulnerability of pocket beaches an assessment using higher resolution spatial data should be applied. The aggregated vulnerability score shows moderate vulnerability while the high vulnerability score for storm surge impact underlines the necessity for short term spatial planning, as is the case in Goksu, Turkey. Although the cliff geomorphology is resistant to flooding, medium cliffs are prone to erosion and the high energy driving force can initiate cliff erosion as well as short term coastal erosion along pocket beaches. The possibility of cliff erosion along the most exposed coast strip is also documented in Lorenzo et al. (2007).

The assessment of the B'Buga region in Malta demonstrates a moderate score of aggregated vulnerability, very similar to that of Viveiro, Spain. Although the scores are close, the ranking of the impact vulnerability scores are different, showing that the aggregated vulnerability score by itself might mask the importance of individual impacts. The same discussion holds true for the different results of the assessment using different spatial resolutions, as with the case of Viveiro. The dominant geomorphology of the B'Buga region is its rocky cliffs, although inside the bay there are small beaches protected by a huge port infrastructure that has recently been constructed. The resilience of the geomorphology combined with milder forces (waves and tides) for geomorphologic processes determines the low to moderate vulnerability in terms of the impacts along the shoreline. Coastal erosion scores signal the low vulnerability of the region, which is also protected from the driving forces via coastal structures. Storm surge and inundation impacts show moderate to low vulnerability. These scores can be attributed to the natural resilience of the coastal strip and significant human intervention which act as protection measures. On the contrary, the vulnerability of groundwater resources for the whole island shows high vulnerability. The over-exploitation of the available resources due to land use and urbanisation, as well as the low resilience of the aquifer, generates the high vulnerability score for this region.

In terms of ICZM planning, the scores show that the most vulnerable resource is groundwater. Although the natural vulnerability of the aquifer is moderate, the high impact of human influence on the vulnerability highlights future problems and urges the preparation of policy actions both for the short term and the long term. The installation of RO plants on the island is one of the actions that can be considered in line with the results of the assessment. However, more actions are needed in the long term. Similar to the case of Viveiro, the assessment of aggregated values for a long coastal strip masks the vulnerabilities of small beaches within the region which might have higher scores. Thus, a similar type of assessment is required using detailed data, the results of which might point out a more definite framework for the future application of ICZM plans.

The vulnerability scores for Goksu point out that the region is highly vulnerable to geomorphology dependent coastal processes such as erosion, flooding and inundation.

Figure 7 shows the influence graphs of Goksu, Turkey in addition to the impact and vulnerability scores. These graphs are important for local decision-making processes, while the comparison of different sites according to the overall vulnerability scores enables planning for the regional to national management of coastal areas. The histogram shows that although human influence on the geomorphologic processes is significant (scores for human influence parameters indicate moderate vulnerability), it is the physical properties of the region that governs the vulnerability. Many of the physical parameters are part of the geomorphologic mechanisms, either as driving forces or as affected attributes, and the scores of these parameters for the Goksu region signal a high vulnerability as reflected by the aggregated vulnerability of the whole region. On the other hand, the vulnerability of groundwater resources is human influence-driven, although the physical characteristics of the aquifers indicate the resilience to sea level rise. It is the establishment of the level of influence of different processes along the coastal area that enables us to generate a framework for the Goksu region in terms of ICZM planning. In terms of high vulnerability impacts where geomorphologic processes govern the dynamics, the understanding of local geomorphology dynamics and the impact of human activities over the long term represents the key areas that ICZM practice should be based on. The high score of flooding due to storm surges indicates that a short term temporal scale should be included in the modelling, underlining the necessity for numerical model studies for this site. Since the physical properties of the region dominate the vulnerability, management options need to be more structure-based, at least in terms of soft protection options such as nourishment or dune planning (Ozyurt & Ergin, 2010).

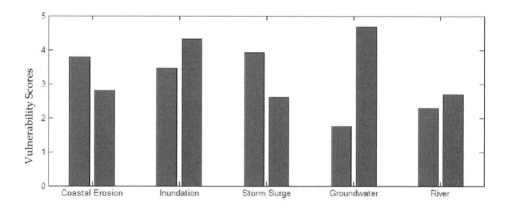

Fig. 7. Influence Histogram for Goksu (red columns indicate the human inference system, blue columns indicate the physical inference system).

As was previously mentioned, ICZM aims to manage all the resources available on the coastal area, including fresh water resources. For Goksu, both river and groundwater resources are assessed in terms of vulnerability to sea water intrusion. Although the river shows moderate vulnerability - due to higher scores of human influence parameters - the possibility of higher vulnerability can be expected in the long term. This result indicates that ICZM should consider adjusting the human activities along the spatial extent of the region, including the river basin management. On the other hand, groundwater resources show high resilience if not over exploited. However - as is shown by the histogram - the human parameters show the highest vulnerability score. Thus ICZM plans must consider policy-driven options that include both short term solutions as well as long term applications for the sustainability of aquifers.

6. Conclusion

For the sustainability of coastal areas, integrated coastal zone management has become the leading concept which requires the integration of many concepts studied by different disciplines, such as geology, geomorphology, coastal and marine sciences, sociology, etc. The study of the geomorphology of coastal areas - focusing on landforms and the processes that shape them - is one of the core disciplines required for successful and efficient ICZM practice. The information generated by geomorphologic studies acts as a foundation for other studies included in ICZM plans - such as vulnerability assessments - by determining the scale of the assessments, the processes to be included and options to be assessed.

The results of the case study locations assessed by the FCVAM are used to discuss the role of geomorphology in the vulnerability of coastal areas. Additionally, the integration of spatial and temporal scales within the model, considering the different scales of geomorphology and the ICZM, are presented through these examples. The assessment methodology uses the concepts and theories of geomorphology (landform processes, drivers and factors) such that different processes (such as waves, tides) acting on the geomorphology are integrated and evaluated by using governing parameters which are not limited to spatial or temporal scales. Geomorphology in terms of coastal land forms is directly included in the FCVAM model. In addition, the site-specific application of the model is suggested to be performed by preparing the model database, determining coastal strips with respect to their geomorphological properties and focusing on landforms. The processes related to specific landforms are the main properties that also define the vulnerability of the coastal zone, and the relationships between these processes and landforms are the structural backbone of FCVAM model. The selection processes of the parameters to be included in site-specific assessments using the FCVAM model are dependent on the study of geomorphology and its theories of specific landforms, as presented in the discussion on the model's parameters.

The fuzzy coastal vulnerability assessment methodology (Ozyurt, 2010) was used to assess different coastal areas showing various physical and geomorphological properties as well as different levels of socio-economic development patterns. The regional application of the FCVAM is presented by comparing three different locations (Viveiro, Spain; B'Buga, Malta and Silifke, Turkey) at the LAU2 level using coarse resolution data. The data used for the regional application of the model needs to represent the properties of the region. However, the coarse resolution of the data used for the FCVAM's application is efficient enough to analyse the vulnerability of these regions and compare them to each other so as to determine regional policies on coastal zone management. On the other hand, the vulnerability scores of individual impacts and the histograms give the most information on the level of vulnerability and the influence of geomorphologic processes of a region. The degree of human intervention on these processes is also presented in the histogram provided in Figure 7. Both the scores and histogram shown in Figure 7 enable policy makers to develop ICZM plans in the long term by creating a framework of possible actions. However, in order to generate efficient histograms, the model needs to be run at a local level with high resolution data or for regions where geomorphology is homogeneous for the study area. Such a case is represented by the study on Goksu, Turkey. The site where geomorphology is dominated by delta formation enables the FCVAM model to analyse relationships between physical and human influence parameters as well as indicating possible adaptation measures for different impacts.

One of the recurring themes is the masking of variability of vulnerability along a shoreline as a result of the application of the model to a coarser spatial resolution. The use of aggregated data to define some of the parameters - especially parameters related to geomorphology - can mask higher or lower vulnerability zones, such as was the case with the pocket beaches in Viveiro. In that case, although the variability is lost in terms of geomorphologic processes, the impact vulnerability scores still help to understand the variability of vulnerability across different types of processes.

Finally - as was previously highlighted - geomorphologic processes are both derived and driven by many mechanisms, and a combination of these mechanisms is the goal of efficient ICZM practice. To achieve this objective, the models of different natures and complexities try to overcome many of the problems faced by geomorphology research and ICZM practice. The FCVAM model and case studies presented represents one of these models and tries to achieve the integration of different processes efficiently. In the end, the problems related to many of the concepts above mentioned are what drive many researchers from many disciplines to continue searching. As Malcolm Muggeridge (What I Believe) has put it in words *"IF I COULD UNDERSTAND A GRAIN OF SAND, I SHOULD UNDERSTAND EVERYTHING."*

7. Acknowledgment

The development of the Fuzzy Coastal Vulnerability Assessment Model is partly supported by research from "Kıyılarda İklim Değişikliğine Karşı Kumlanma Modeli Destekli Kırılganlık Analizi Projesi – KIDEKA" (Coastal vulnerability assessment model coupled with sediment transport model) Project supported by the TUBITAK Research Grant No: 108M589.

8. References

Axiak, V. (1992) Implications of expected climatic changes on the island of Malta: identification and assessment of possible climatic change on marine and freshwater ecosystems. UNEP Report.

Beatley, T., Brower, D.J. and Schwab A.K. (2002). "An Introduction to Coastal Zone Management". Second Edition. Island Press.

Bird, E. C. F. (2008). "Coastal geomorphology : an introduction" Second Edition. Wiley.

Birdi, N. (1997). Water scarcity in Malta. *GeoJournal*, 41.2: pp. 181–191.

Davis, Richard A. (1993) "The Evolving Coast." Scientific American Library. Printed in USA. p. 231.

French, J.R. and Burningham, H. (2009). Coastal geomorphology: trends and challenges. *Progress in Physical Geography*, Vol 33 No. 1 pp. 117–129.

IPCC. (2007). "Intergovernmental Panel on Climate Change (IPCC) Fourth Assessment Report: Climate Change 2007 Impacts, Adaptation and Vulnerability." M.L. Parry, O.F. Canziani, J.P. Palutikof, P. J. v. d. Linden, and C. E. Hanson, eds., IPCC, Cambridge, United Kingdom and New York, NY, USA.

Jiongxin, X., 2004. A study of anthropogenic seasonal rivers in China. *Catena*, Vol. 55 pp. 17–32.

Klein, R. J. T., & Nicholls, R. J. (1999). Assessment of coastal vulnerability to climate change. *Ambio*, 28(2), pp. 182–187.

Lorenzo F., Alonso A., Pagés J. L. (2007) Erosion and Accretion of Beach and Spit Systems in Northwest Spain: A Response to Human Activity. *Journal of Coastal Research*, Vol. 23, No. 4 pp. 834-845.

McFadden, L., Nicholls, R. J. and Penning-Rowsell E. (Eds.). (2007). "Managing Coastal Vulnerability". Elsevier.

Nilsson, C., Reidy, C. A., Dynesius, M., and Revenga, C. (2005). "Fragmentation and Flow Regulation of the World's Large River Systems." *Science*, (308), p. 405.

Ozyurt, G. (2007). "Vulnerability of Coastal Areas to Sea Level Rise: Case Study on Goksu Delta," Master's Thesis. Middle East Technical University, Ankara.

Ozyurt,G. and Ergin, A.(2009). "Application of Sea Level Rise Vulnerability Assessment Model to Selected Coastal Areas of Turkey". *Journal of Coastal Research*, Special Edition 56, pp. 248-251.

Ozyurt, G., and Ergin, A. (2010). "Improving Coastal Vulnerability Assessments to Sea Level Rise: A New Indicator Based Methodology for Decision Makers." *Journal of Coastal Research*, 26(2), pp. 265–273.

Ozyurt,G. (2010) "Fuzzy Coastal Vulnerability Assessment to Sea Level Rise" PhD. Thesis. Middle East Technical University, Ankara.

Ospina-Alvarez, N., Prego R., Alvarez I., deCastro M., Alvarez-Ossorio M.T., Pazos Y., Campos M.J., Bernardez P., Garcia-Soto C., Gomez-Gesteira M., Varela M. (2010) Oceanographical patterns during a summer upwelling–downwelling event in the Northern Galician Rias: Comparison with the whole Ria system (NW of Iberian Peninsula) *Continental Shelf Research*, Vol.30 pp. 1362–1372.

Rotmans, J. and Rothman D. S. (2003) "Scaling in Integrated Assessment". Swets & Zeitlinger.

Sanders M. H and Clark P. D. (Eds.) (2010) "Geomorphology: Processes, Taxonomy and Applications". *Nova*.

Schwartz, M. L. (Eds.) (2005) "Encyclopedia of Coastal Science". Springer.

US Army Corps of Engineers. (2003). "Coastal Engineering Manual." Washington, DC.

Woodroffe, C.D. (2002) "Coasts: form, process and evolution". Printed in UK, University Press, Cambridge.

7

Hydro-Geomorphic Classification and Potential Vegetation Mapping for Upper Mississippi River Bottomland Restoration

Charles H. Theiling[1], E. Arthur Bettis[2] and Mickey E. Heitmeyer[3]

[1]U.S. Army Corps of Engineers, Rock Island District, Rock Island, Illinois,
[2]University of Iowa, Department of Geoscience, Iowa City, Iowa,
[3]Greenbrier Wetland Services, Advance, Missouri,
USA

1. Introduction

Ecosystem restoration that incorporates process and function has become well known among ecosystem restoration practitioners (Society for Ecological Restoration, 2004; Palmer et al., 2005; Kondolf et al., 2006;). It has been recommended for the Upper Mississippi River System (UMRS; Figure 1) by expert advisory panels (Lubinski and Barko, 2003; Barko et al., 2006) and in Federal policy (U.S. Water Resources Development Act 2007, Section 8001). Our conceptual model for the UMRS integrates process and function among five Essential Ecosystem Components (EECs; Harwell et al., 1999), with hydrology, geomorphology, and biogeochemistry strongly influencing habitat and biota (Lubinski and Barko, 2003; Jacobsen, in press). The primary ecological driver of large floodplain river landscapes is hydrology (Junk et al., 1989; Poff et al., 1997; Sparks et al., 1998; Whited et al., 2007; Klimas et al., 2009), with discharge and river stage being the most common indicators of system condition and variability. Hydrology and hydraulics are conditioned by the geomorphic setting, or geomorphic landscape, which establishes river stage and floodplain inundation response to variable discharge (Clarke, et al., 2003; Thoms, 2003; Newson, 2006; Stallins, 2006; Thorp et al., 2008). Geomorphology is frequently presented as planform aquatic features (i.e., channel, secondary channel, backwater, floodplain, etc.), the river cross-section, floodplain topography, or soil profiles and maps. Flood inundation patterns are mapped less frequently, but they are strongly influenced by both regional and local hydrology and geomorphology (Thorp et al., 2008).

The UMRS is an institutional designation that includes the Upper Mississippi River Valley (UMV), the Illinois River Valley (IRV) and small parts of several tributaries (U.S. Water Resources Development Act 1986, Section 1103) which together span about 1,200 miles of 9-foot deep channels (Figure 1; USACE, 2004a). Channel clearing and stabilization under Federal authority began in 1824 and culminated with 37 lock and dam sites and thousands of channel training structures (USACE, 2004a). Chronic and sporadic shoaling requires dredging every year despite construction of low head navigation dams and channel regulating structures.

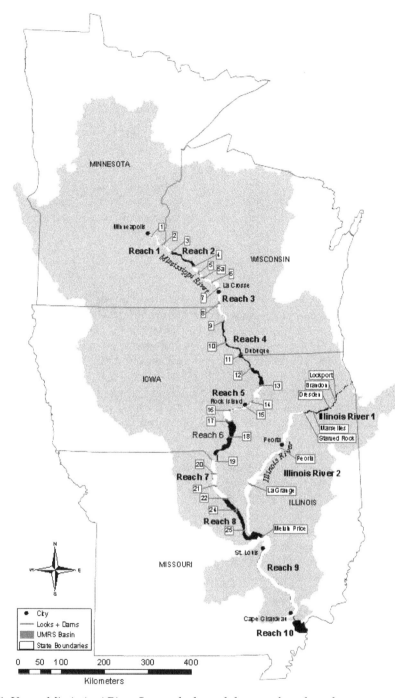

Fig. 1. Upper Mississippi River System locks and dams and pool reaches.

The entire river-floodplain covers more than 2.6 million acres (Theiling et al, 2000). The river includes four large "floodplain reaches" (Figure 2) defined by large scale valley features and social impacts (Lubinski, 1999). There are also 18 "geomorphic reaches" which were defined using riverbed slope, valley and channel features, and tributary confluences (WEST Consultants, Inc, 2000; Theiling, 2010). Geomorphic reach characteristics are important determinants of environmental response to development and floodplain land use objectives. Floodplain reaches and geomorphic reaches are analogous with Functional Process Zones and River Reaches, respectively, defined by Thorp et al. (2006, 2008) in their River Ecosystem Synthesis.

Floodplain development occurred concurrent with European settlement and industrialization. Increased shipping demand and the introduction of steamboats consumed massive amounts of wood from the floodplain (Norris, 1997) and necessitated channel improvements to carry larger loads during low flow periods and droughts. When forests were cleared for fuelwood and lumber, agriculture moved in to exploit the rich alluvial environment. Individual farmers connected natural levees to increase crop success initially and later constructed formal levee and drainage systems (Thompson, 2002).

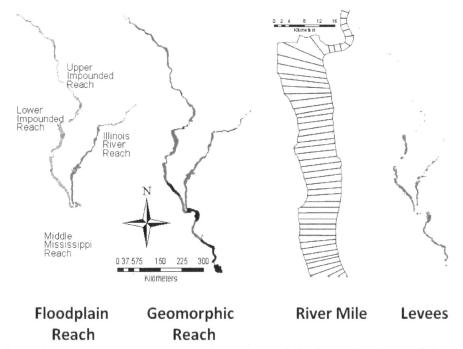

Floodplain   Geomorphic   River Mile   Levees
Reach   Reach

Fig. 2. Upper Mississippi River geomorphic scaling includes large, glacial controlled floodplain reaches, fluvial geomorphic controlled reaches, and structured (river mile) or political (levee district) segmentation schemes.

Levees (Thompson, 2002; USACE, 2006), water diversions (Starrett, 1971), and dams (Chen and Simmons, 1986; Fremling et al., 1989) were completed at system-wide scales to manage the distribution and conveyance of surface waters to control flooding, dilute municipal pollution, support navigation, and enhance habitat. The outcomes of these changes differ depending on location in the river system (Theiling and Nestler, 2010).

Alterations to hydrology, geomorphic structure, and direct impacts from historical land use change have substantially altered the form and function of ecological communities and processes in the UMRS. The flow of energy is a critical function in ecosystems and alterations to energy pathways can cascade through ecosystems in many ways (Welcomme, 1979; Vannote et al., 1980, Ward and Stanford, 1983; Junk et al., 1989; Ward et al., 1989). Early formal models for stream ecosystem energetics emphasized linear pathways transporting and utilizing metabolic energy differently along a river continuum (Vannote et al., 1983). The early stream ecosystem conceptual models were then tailored to account for nutrient cycling (Newbold et al., 1982), anthropogenic disturbances (Ward et al., 1999), different types of rivers (Junk et al., 1989; Wiley and Osborne, 1990), internal processes (Thorp et al., 1994), watershed influence (Benda et al., 2004), and geomorphic structure (Thoms, 2003). We developed system scale data to focus on the relationships expressed in the hydrogeomorphic methodology (Brinson, 1993; Klimas et al., 2009) and the River Ecosystem Synthesis (Thorp et al., 2006, 2008). Land cover, aquatic area, hydrology, and geomorphology data were derived for the entire UMRS for historic, contemporary, and simulated conditions. They can be compared among functional units such as Functional Process Zones or Hydrogeomorphic Patches defined by Thorp et al. (2008) or reference conditions (Nestler et al., 2010; Theiling and Nestler, 2010; SER, 2004) to create simulations of potential vegetation communities under alternative management scenarios.

Ecosystem restoration initiatives require estimates of the natural resource benefits that may be achieved by alternative project plans or project features to ensure accountability and success in Federal projects (USACE, 2000). Recent guidance also calls for the use of adaptive management in Federal water resource planning (Section 2039 of U.S. Water Resources Development Act 2007; Council on Environmental Quality, 2009). The models described here are important elements of adaptive management because they can estimate anticipated outcomes for comparison during monitoring and evaluation stages of the adaptive management cycle (Christianson et al., 1996; Walters, 1997; Williams, 2009). Many restoration plans include plant community or habitat models that estimate community response to physical forces (U.S. Water Resources Council, 1983; USACE, 2000; Council on Environmental Quality, 2009). Predicted plant and habitat response can then be used to support species or community habitat suitability models (USFWS, 1980). Dynamic physical forces are well known ecological drivers in large rivers (Doyle et al. 2005). Methods and data presented here can help estimate physical-ecological cascades resulting from hydrologic and geomorphic alteration of large rivers. We have made great progress developing data needed for potential vegetation models for the entire system. We also discuss the need for a rigorous landscape analysis that includes forest composition in the pre-settlement land cover data.

2. Methods

2.1 Geomorphology

Riverbed slope, channel geometry, and substrates are well known for engineering purposes. System-wide topographic mapping and channel surveys undertaken for each significant channel improvement plan were completed in 1890 and 1930. Surveys are much more frequent in the modern era. The Valley's floodplain has been mapped to document the relative age of geomorphic surfaces and associated deposits to help manage cultural resources (Bettis et al., 1996). The studies developed Landform Sediment Assemblages (LSA) which are mappable landforms and their underlying deposits that occur with predictable characteristics (Figure 3; Hajic, 2000). U.S. Department of Agriculture (USDA) soil maps are widely available, but generally lack detail in frequently flooded parts of the floodplain.

Geomorphic mapping in the Valley generally followed the protocol defined by Bettis et al. (1996) with slight variations. U.S. Geological Survey topographic quadrangle maps, aerial photos, soils maps, boring records, and literature were used to construct geomorphic maps. Geomorphic classifications were done at several different scales which allows for more detailed site-specific analysis than reported here. Mapping under modern aquatic areas was not possible and most of the low elevation features (active floodplain and some paleo-floodplain) were inundated in the lower ends of navigation pools 2 through 13 between Minneapolis, Minnesota and Clinton, Iowa. We unioned four separate LSA data sets (Bettis et al., 1996; Madigan and Schirmer, 1998; and Hajic, 2000) and reclassified them using a common classification scheme in GIS. The data were clipped to the bluff to bluff floodplain extent (Laustrup and Lowenburg, 1994). LSAs were summarized using a river mile segmentation floodplain overlay. River mile segments are unequal because the width of the floodplain varies and there are curves in the river that create wedge-shaped polygons. These results are a first approximation and open to further interpretation. Higher resolution mapping and analysis will be required for site-specific studies (Heitmeyer, 2010), but this generalized classification matches flood inundation mapping, historic land cover mapping, and regional habitat assessments (Theiling et al., 2000) quite well.

Our LSA geomorphic classification has nine classes described below. Characteristics were derived from Bettis et al. (1996), Madigan and Schirmer (1998), and Hajic (2000) and mapped as follows:

- **Modern Aquatic Classes** (Modern Channel, Modern Backwater) are primarily the result of navigation dams that inundated low elevation active and paleo-floodplain geomorphic classes, leaving levees and ridges exposed as islands in impounded aquatic areas in Pools 2 to 13. Aquatic area is generally <10 percent of the total floodplain area south of Rock Island, Illinois, but 20 to 60 percent in the north upstream from Rock Island. Modern aquatic area ranges from a few hundred to over 1,800 acres per river mile. Aquatic area is <500 acres for most river mile segments except at Illinois River miles where large lakes occur and on the Mississippi River where impoundment effects are exhibited in Pools 3 to 13.
- **Active Floodplain – Poorly Drained** is low elevation floodplain of vertical accretion origin that would have been or is flooded most years. These areas are often associated with tributary confluences. Soils are likely silt, loam, clay mixes that grade downward to coarser sand and pebbly sand. Fine sediments may be 1 – 2 meters deep over coarser sediment. These surfaces are inundated in the lower portions of all navigation pools. Some of these areas occur riverward of the flood control levees where they are exposed to altered hydrology and material transport. Similar areas behind levees are isolated from the river and may maintain more of their historic characteristics. Active floodplain is most abundant in the mid valley Mississippi River reaches and lower Illinois River. That is likely due to the limited effects from impoundment and the drainage of low elevation floodplain in agricultural drainage districts.

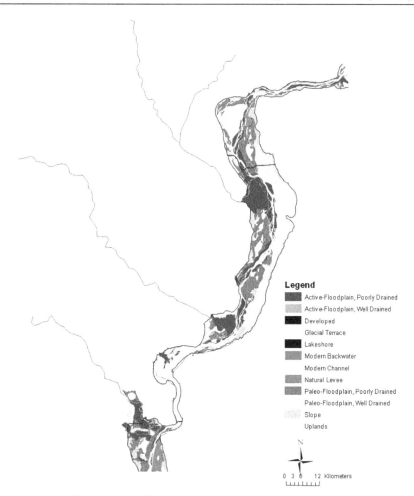

Fig. 3. Landform Sediment Assemblage maps characterize surficial and underlying characteristics that help define local edaphic factors. This map depicts parts of Pools 16 to 20 between Muscatine, Iowa and the junction of the Des Moines River.

- **Active Floodplain – Well Drained** is frequently flooded low elevation floodplain of lateral accretion origin. It is underlain by less than 1.5 meters of fine-grained alluvium that buries sand and pebbly sand. Despite high frequency inundation, it does not retain water. Dry active floodplain may also be associated with alluvial fans and deltas. Dry active floodplain is common on the Illinois River and occurs in patches in the St. Paul District. This class was not mapped in the Rock Island and St. Louis Districts.
- **Paleo-Floodplain – Poorly Drained** is infrequently flooded mid elevation floodplain of vertical accretion origin. These floodplain areas contain former channel and lake features that have transitioned to terrestrial area. Deposits and soils are variable with fine silt, loams, and clays overlying pebbly sand. They function as overflow channels on the rising and receding flood or as ponded groundwater at high river stage. They

formed backwater lakes and sloughs prior to significant floodplain drainage (Heitmeyer, 2008).

- **Paleo-Floodplain – Well Drained** is infrequently flooded mid elevation floodplain of lateral accretion origin that includes inactive scrolls, bars, meander belts, and splays. Soils are variable with fine silt, loams, and clays overlying sand and pebbly sand. Paleo-Floodplain is mapped mostly in the Rock Island and St. Louis Districts. In the Rock Island District it is an association with early and mid Holocene surfaces that define the wet areas and paleo-channels that derive the dry areas. In the St. Louis District this LSA comprises large meander scrolls that occupy a major proportion of the more elevated floodplain area. There is almost no paleo-floodplain in the St. Paul District because Holocene channel incision has isolated older surfaces as infrequently flooded terraces. Older surfaces in the St. Paul District occur as terraces.

- **Natural Levees** are slightly elevated, well-drained areas that parallel relatively stable channel reaches. Levees may also occur at crevasse splays that extend from channels cut into the natural levee and spreading into adjacent low-lying wet paleo-floodplain. Deposits of this LSA are stratified loam, sand, silt, clay, and sand. Levees are discontinuous linear areas that appear most abundant on the Illinois River because the Illinois River mapping was done at a smaller scale (higher resolution; Hajic, 1990). Several large levee areas are mapped in the Rock Island District and smaller levee areas are common along the channel in the St. Paul District where they are not submerged.

- **Alluvial/Colluvial Aprons** are elevated, bluff-base areas underlain by a variety of sediments derived from adjacent slopes and small tributary valleys. This LSA typically is quite messic and is rarely inundated. The most notable abundance of this LSA occurs in Illinois near Quincy where there are other high floodplain features.

- **Sandy Terraces** occur throughout the river and were formed during the last glacial period (Knox and Schumm in West Consultants, Inc., 2000). They are most abundant in the Illinois, Minnesota, Chippewa, Maquoketa, and Iowa River reaches. Downstream of the Iowa River Reach this LSA merges with the paleofloodplain LSA.

2.2 Hydrology

High resolution topographic data and updated river stage-discharge relationships were developed following the "Great flood of 1993" when there was a comprehensive review of floodplain management (Interagency Floodplain Management Review Committee, 1994). Photogrammetric methods were used to create a high accuracy digital elevation model for the entire Upper Mississippi floodplain for use in hydrologic modelling to re-define the river stage frequency rating curves. We created GIS overlays of the water surface elevation profiles corresponding to the rating curves, superimposed on the high resolution topography to map potential flood inundation patterns (Figure 4) for 8 annual exceedance probability floods: 50, 20, 10, 4, 2, 1, 0.5 and 0.2 percent (i.e., 2-, 5-, 10-, 25-, 50-, 100-, 200-, and 500-yr expected recurrence interval flood).

2.2.1 Topographic data

The U.S. Geological Survey National Elevation Database available through the National Map Seamless Server provided online access to digital elevation data in an easily accessible and well documented format. Upper Mississippi, Illinois, and Missouri Rivers floodplain elevation data were updated in 1998 using high resolution stereographic techniques (Interagency Floodplain Management Review Committee, 1994; Scientific Assessment and Strategy Team (SAST), David Greenlee, USGS EROS Data Center, Sioux Falls, South Dakota, personal communication). The Mississippi River floodplain ("bluff-to-bluff") digital terrain model data was designed and compiled so that spot elevations on well-defined features would be within 0.67 feet (vertical) of the true position (as determined by a higher order method of measurement) 67% of the time. It is approximately 1/6th of a contour interval (4 foot contours; U.S. Army Corps of Engineers, 2003, 2004b). High river stages when photography was acquired limited their utility to visualize and model low river stages in mid reaches of the Mississippi River and prevented their use for this project on the Lower Illinois River. The NED2003 floodplain elevation data were used for the Illinois River floodplain inundation mapping. Issues regarding vertical datum conversions were evaluated and determined to be insignificant at the scale and intended application for this study (Theiling, 2010).

Data can be accessed at several levels of resolution, we used the default 1 arc second download format to conserve data processing requirements over large geographic regions and because subsequent hydrologic modeling analyses were completed at similar resolution. Rectangular tiles covering about 100 miles each were downloaded and data extracted by a mask of the floodplain as represented by the prior defined floodplain extent for each pool (Laustrup and Lowenburg, 1994). We combined the pool scale DEMs into a DEM for the entire floodplain using default mosaic procedures in ArcGIS. Metric elevations were converted (i.e., times 3.281 in Raster math) to English units to match river stage in feet and discharge in cubic feet per second (cfs) which is the vernacular of the Flow Frequency Study.

2.2.2 Flow frequency study

Hydrologic analyses were accomplished with 100 years of record from 1898 to 1998 using the log-Pearson Type III distribution for unregulated flows at gages. Mainstem flows between gages were determined by interpolation of the mean and the standard deviation for the annual flow distribution based on drainage area in conjunction with a regional skew. Flood control reservoir project impacts were defined by developing regulated versus nonregulated relationships for discharges, extreme events were determined by factoring up major historic events, and the UNET unsteady flow program was used to address hydraulic impacts. The result of the hydrologic aspects of the study was a discharge and related frequency of occurrence for stations or given cross sections located along the Mississippi and Illinois Rivers (Figure 4; USACE, 2004b).

A hydraulic analysis was required to establish the water surface elevation associated with each frequency of discharge at each location or cross section along the river reach. The main procedures were to use the UNET unsteady flow numerical modeling tool with recent channel hydrographic surveys (routinely obtained for navigation channel maintenance), and floodplain digital terrain data collected in 1995 and 1998. Levee overtopping was established at the top of existing levee grade based on an upstream and a downstream point. Using these station rating curves and the station frequency flows developed during the hydrology phase, frequency elevation points were obtained for each cross section location. Connecting the corresponding points resulted in flood frequency elevation profiles (USACE, 2003).

2.2.3 Floodplain inundation

Triangulated Irregular Network (TIN) files were created from the cross section feature lines for each separate flood stage frequency (Figure 4). Each flood stage TIN: 50%, 20%, 10%, 4%, 2%, 1%, 0.5%, and 0.2% annual exceedance probability (i.e., 2-yr, 5-yr, 10-yr, 25-yr, 50-yr, 100-yr, 200-yr, 500-yr expected recurrence interval flood) was overlayed in a cut-fill analysis on the high resolution floodplain topography for each navigation pool or reach. The area represented as inundated by the cut-fill procedure for each flood stage was separated out as a conditional GRID analysis that selected areas with volume > 0 and output a single GRID with a count of the 20X20 m cells below the elevation of the water surface elevation (Figure 4). This value was exported to a spreadsheet where grid counts were converted to area estimates (acres) for the navigation pool scale at which they were created. The resulting GRID was converted to a shapefile to merge with other layers to create system-scale layers of the potential water distribution at each flood stage (Figure 4).

Floodplain inundation classes (i.e., 50, 20, 10, 5, and 1 percent annual exceedance probability floods) were summarized by river mile and compared by geomorphic reach. Leveed areas were then extracted from inundation layers to assess changes in flood distribution attributable to levees. These data were also summarized by river mile and geomorphic reach. The inundation classes in leveed areas were subtracted from the maximum simulated inundation surface in each geomorphic reach (i.e., 1 percent or 0.02 percent annual exceedance probability flood) and data were normalized as percent of maximum inundation area.

2.3 Land cover

2.3.1 Presettlement land cover

Land cover databases are the foundation of our vision of UMRS landscapes and habitats over multiple reference conditions. Early explorers described interesting new landscapes, vast abundances of strange new animals, and drew crude maps as they moved through North America (Carlander, 1954). As settlers followed explorers, the Public Land Survey (PLS) mapped and characterized the mostly unsettled Louisiana Territories to sell land to the westward-expanding population of the United States (Sickley and Mladenoff, 2007). The PLS methods first divided the region into 36 square mile townships and then subdivided each one into 36 one mile square sections. Along the township and section lines, the surveyors set posts every half mile at locations called ½ section corners (where section lines intersected) and quarter section corners (midway between the section corners). Between two and four bearing trees were marked near each post and recorded in their notebooks by species, diameter, and compass bearing and distance from the post. The surveyors recorded other features that they encountered along the survey lines in the notebooks as well, including water features, individual trees located between the survey posts, boundaries between the ecosystems through which they were traveling, boundaries of natural and anthropogenic disturbances, and cultural features such as houses, cultivated fields, roads, and towns. Initial pilot studies reconstructing PLS surveys in the UMRS (Nelson et al., 1996) proved to be very valuable, so The Nature Conservancy's Great River Partnership contracted the University of Wisconsin Forest Ecology Lab to complete a comprehensive interpretation in a GIS for the entire UMRS (Sickley and Mladenoff, 2007). PLS data extend beyond the bluff into upland habitats, but the data were clipped to the bluff to bluff extent for this initial analysis. The Nature Conservancy dataset, and recently available statewide PLS plat map GIS coverages, provide a snapshot to speculate on ecological community associations in the undeveloped landscape.

Scale and resolution are important issues to consider when using PLS data. The quarter section and ½ section corners are a half mile apart and are generally marked by two to four trees each. A single section is commonly bounded by eight corners, which means that a square mile in the data would contain information on about only 16 to 32 trees. This is too sparse to be used at a stand or site level in anything other than the most qualitative sense. It is recommended to use the data at broad spatial extents (tens to thousands of square miles) and at resolutions of no less than a square mile (Schulte and Mladenoff, 2001; Theiling, 2010).

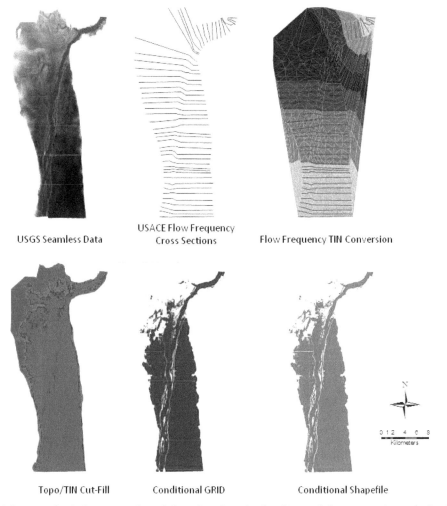

Fig. 4. Images depicting examples of elevation data, hydraulic model cross-sections, derived TINs, cut/fill interpolation, and grid and shapefile products.

2.3.2 Contemporary land cover

Environmental Management Program Long Term Resource Monitoring (LTRM) has compiled several system-wide land cover data sets. The 2000 land cover data extent was used to define the floodplain area for other GIS coverages. LTRMP Land cover data were interpreted from 1:15,000 scale infra-red aerial photography with a minimum map unit of one acre. Several land cover classifications schemes have been used, but National spatial data standards have helped optimize and standardize the scheme. The current classification scheme includes 31 classes that are ecologically or socially relevant. The scheme can be lumped or split as necessary to match other data sets. The HNA-18 land cover classification was reclassified to the general ecosystem classes compatible with the PLS data (Theiling, 2010). LTRM land cover data were combined in a spatial join to replicate the point sampling scheme of the PLS on the contemporary data.

2.3.3 Land cover classes

Land cover data from historic and contemporary periods were generalized to a common 12 class scheme (Theiling, 2010). The classification scheme combined several forest classes from the contemporary classification and two from the historic classification. The savanna class combined 11 classes from the PLS surveys, but none from the modern surveys because the habitat is only rarely present in the modern landscape. A "bottom" class was evident in the historic data but not clear in the contemporary data which were lumped as "forest." Similar to forests, the historic data allowed separation of several prairie classes: prairie, bottom prairie, and wet prairie which were not separable in the modern data. The historic classification identified forested wetlands as swamps, but that distinction is not made in the contemporary data where forested wetlands were not identified. Shrubs were represented in both data sets. Water was classified as several aquatic area types in the historic data, but in the modern data distinctions among aquatic classes depended on the presence of vegetation. Agriculture and developed classes were not common in the historic data, but they were very important in the modern data. PLS data have been criticized for inaccurate and inconsistent identifications and naming conventions. Their use at the general landscape level here is to provide a broad view of the system without consideration of species and precise locational information.

2.4 Data analysis

We overlayed the river reach segmentation schemes on land cover layers to provide proportional estimates for each land cover class to show plant community composition change along the river. A GIS extension was built to complete point counts for each land cover class at each river mile (Tim Fox, U.S. Geological Survey, Upper Midwest Environmental Sciences Center, La Crosse, Wisconsin). We also summed point counts by geomorphic class and hydraulic inundation frequency. Data were normalized as a proportion of total points within each segment (i.e., river mile, pool, reach, etc.) to assess the relative importance of each class in each area. The normalized data were plotted by river mile here and also used in multivariate statistical analyses examining the distribution of geomorphic, hydrologic, and land cover characteristics among river reaches at several scales (Theiling, 2010).

2.5 Hydro-geomorphic methodology

The HGM process of evaluating ecosystem restoration and management options relies heavily on eight types of data, most of which require geospatial digital information usable in an ArcGIS/ArcMAP format. These data include historic and current information about: 1) soils, 2) geomorphology, 3) topography/elevation, 4) hydrology/flood frequency, 5) aerial photographs and cartography maps, 6) land cover and vegetation communities, 7) presence and distribution of key plant and animal species, and 8) physical anthropogenic features.

The three-stages of HGM are as follows: first, the historic condition and ecological processes of an area and its surrounding landscapes are determined from a variety of historical and current information such as geological, hydrological, and botanical maps and data. Public Land Survey (PLS) maps and notes are especially useful to understand historic vegetation composition and distribution. A key element of HGM is developing a "matrix" of understanding of which plant communities historically occurred in different geomorphological, soil, topographic, and flood-frequency settings (Table 1). For example, in the Mississippi-Missouri River Confluence Area, wet bottomland prairie that was dominated by prairie cordgrass historically occurred at elevations greater than 417 feet, on relict alluvial floodplain terrace surfaces, on silt loam soils, and between the two- and five-year flood frequency zones (Heitmeyer and Westphall, 2007). Contemporary areas that offer these conditions, especially surface, soil, and flood frequency attributes now offer the best edaphic conditions for restoring wet bottomland prairie communities.

Second, alterations in hydrological condition, topography, vegetation community structure and distribution, and resource availability to key fish and wildlife species are determined by comparing historic vs. current landscapes. This analyses is essentially a qualitative "best professional judgment" assessment of current condition and the types and magnitudes of changes, including assessment of which communities are most resilient and which types of change are the most/least reversible.

Third, options and approaches are identified to restore specific habitats and ecological conditions. The foundation of ecological history coupled with assessment of current conditions helps to determine which system processes (e.g., periodic dormant season flooding) and habitats (e.g., forest composition) can be restored or enhanced, and where this is possible, if it is at all. Obviously, some landscape changes are more permanent and less reversible (e.g., mainstem levees on the Mississippi and Illinois rivers) than others (e.g., clearing of bottomland forest). Through development of the HGM matrix conservation planners can identify: 1) which, and where, habitat types have been lost or altered the most and establish some sense of priority for restoration efforts; 2) where opportunities exist to restore habitats in appropriate geomorphic, soil, hydrological, topographic settings including both public and private lands; 3) how restoration can replace lost functions and values including system connectivity; and 4) what management types and intensity will be needed to sustain restored communities. HGM can be an iterative process that is well-coupled with adaptive ecosystem management (Christensen et al., 1996; Palmer et al., 2005) because new monitoring and research can be used to refine HGM models and restoration plans.

Habitat Type	Geomorphic Surface	Soil Type	Flood Frequency
Open Water	Active river channels, side channels	Riverine Riverine	Permanent Permanent-seasonally dry
	Abandoned channels	Clay, silt-clay	Permanent-seasonally dry
Bottomland Lake	Abandoned channels	Clay, silt-clay with sand/loam plugs	Permanent to semi-permanent
Riverfront Forest	Bar-and-chute and braided bar	Sand, sandy loam and silt loam in swales	1 – 2 year
Floodplain Forest			
Ridges	Point bar ridge	Loam, sandy loam	2 – 5 year
Swales	Point bar swales and tributary riparian zones	Silt loam, slit clay veneer	1 – 2 year
Bottomland hardwood forest	Backswamp, larger point bar swales and floodplain depressions	Silt loam, silty clay	2 – 5 year
Slope Forest	Alluvial fans, colluvial aprons, terrace edges	Mixed erosional	>20 year
Savanna	Alluvial fans, colluvial aprons, terrace interface	Silt loam	10 – 20 year
Bottomland Prairie			
Wet	Point bar and terrace swales and depressions	Clay, silt clay	2 – 5 year
Intermediate	Point bar ridges	Silt loam	>5 year
Mesic Prairie	Point bar edges and terraces	Sandy loam, silt loam	>20 year

Table 1. Hydrogeomorphic matrix of historic distribution of major vegetation communities/habitat types in the American Bottoms geomorphic reach (near St. Louis, Missouri) in relation to geomorphic surface, soils, and flood frequency.

3. Results

3.1 Gemorphology

Land Sediment Assemblage abundance plotted by rive mile illustrates the distribution of each class and the relative width of the floodplain (Figure 5, top). Geomorphic reach overlays helped identify characteristics that separated reaches in a multivariate analysis (Theiling, 2010). The Chippewa River Reach (RM 650 – 750) is separated downstream by the narrower Wisconsin River Reach (RM605-650) which runs through resistant dolomite valley walls (Knox, 2007). The floodplain widens again through erosive shale in the Maquoketa

River Reach (RM510-605) to the Rock Island Gorge (RM465-510) which presents another constrained, resistant dolomite reach (Trowbridge, 1959). Significant widening occurs just below the gorge where the Mississippi Valley intersects an ancient bedrock channel (Iowa River Reach RM420-465). Sandy terraces are abundant in the Iowa Reach and broader reaches upstream (Figure 5, bottom), but they are buried below Holocene sediments downstream of Quincy Illinois near river mile 325. Alluvial/Colluvial apron is ubiquitous, but uniquely abundant in the Des Moines River, Quincy Anabranch, and Sny Anabranch Reaches (RM240-400) where perched wetlands were once present. Paleofloodplain created from Missouri River outwash in the early Holocene is the dominant LSA class at the confluence with and south of the Missouri River (RM200; Bettis et al., 2008). Active floodplain abundance and distribution is relatively constant among reaches. The abundance of aquatic area is higher upstream from river mile 400 because of the effect of dams increasing surface water area in a series of shallow navigation pools (Theiling and Nestler, 2010).

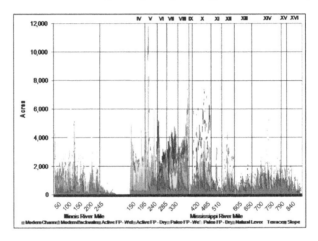

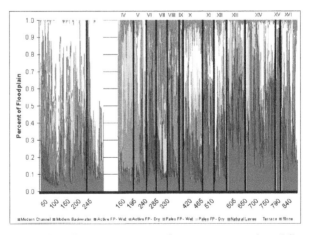

Fig. 5. Geomorphic class distribution in acres and as proportion of total floodplain area for the Upper Mississippi River System.

The Illinois River floodplain presents a diverse land sediment assemblage (Figure 5; Hajik, 2000). The Upper Illinois (> river mile 245) is deeply flooded by dams and only Sandy Terraces remain visible. The Lower Illinois River has not been subdivided into reaches here, but other authors have defined three or more reaches (Starrett, 1972; Sparks, 1992). Terraces are the most abundant floodplain feature, but natural levees are also widely distributed. Active floodplain surfaces increase at river mile 100 below the confluence with the Sanganois River, a major tributary. The Lower Illinois River is slightly narrower than the Mississippi (Figure 5) and it has a much lower gradient than most rivers (Starrett, 1972).

3.2 Hydrology

The abundance of water mapped at low flow periods was relatively constant in the river in 1890 (Figure 6, bottom). Several large aquatic areas: Lake Pepin – River Mile 765, Lima Lake – River Mile 350, MMR Backwaters <River Mile 200, were notable features of the floodplain in 1890, but now only Lake Pepin and degraded and disconnected MMR backwaters persist. The contemporary distribution of surface water (Figure 6) reflects the impact of navigation dams completed ~1940 (Theiling and Nestler, 2010). Water surface area increases in impounded reaches upstream of RM400 and a repeating pattern of dam effects are apparent. UMRS navigation dams are only required to maintain low flow navigation, and their impoundment effect only extends partway up each navigation pool (Theiling and Nestler, 2010). Dam gates are raised out of the river during flood stage, except at Dam 19 (hydropower), about 15 percent to 50 percent of the time (USACE, 2004c, 2004a) when discharge alone can maintain navigable depths.

The change in distribution of aquatic classes is quite striking in the floodplain upstream from the Rock Island Gorge (~River Mile 500) where impoundment effects are pronounced (Figure 7). Sandbars were lost throughout the river system coincident with increased river stages. Wooded islands were lost in the upper river reaches during the post-dam era because of wind-wave erosion of former floodplain ridges and levees exposed following impoundment (Rohweder et al., 2008). The increase in contiguous, or connected, backwaters is a very prominent change in the upstream reaches, but not very important in lower reaches. Isolated backwaters were not prominent in either period, but they are considered very important for many flora and fauna.

Floodplain inundation differs throughout river valleys in response to many natural and anthropogenic drivers. Major tributary rivers demark most geomorphic reaches and each contributes flow and its unique sediment signature to the mainstem Mississippi and Illinois Rivers. The wider banded segments in Figure 8 (bottom) represent areas of greater floodplain inundation diversity which typically occurred at tributary fans and in steep valley reaches. Areas where all the flood stages are compressed (e.g., below river mile 125) are primarily influenced by frequent floods that would fill most of the valley. The impact of the navigation system is apparent in the amount that "Pool Stage" increases as a proportion of maximum inundated area upstream from river mile 400. The distribution of the 2-year flood is prominent along the entire river where it commonly exceeds 70 percent of the total floodplain area and 90 percent in a few locations. This is a characteristic of floodwater distribution across a range of streams and rivers (Leopold et al., 1964).

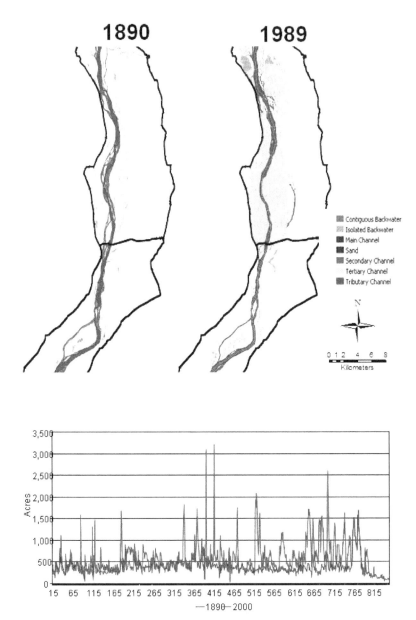

Fig. 6. Surface water impacts from impoundment differ in the northern and southern parts of the system as represented by acres of surface water (bottom) and the map of the Lock and Dam 13 area at River Mile 522. Dam effects in the upper pools are similar to the upper portion of the 1989 image with large contiguous backwaters created by dams, whereas dam effects in downstream pools are more similar to their pre-dam form as shown in the bottom part of the 1989 image.

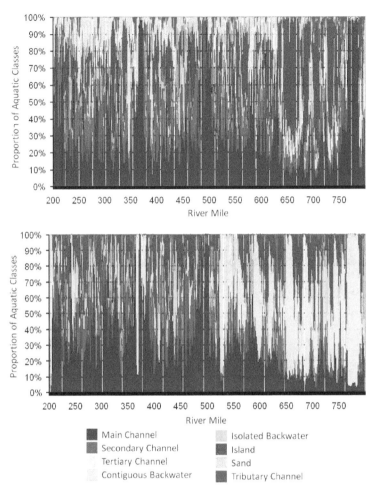

Fig. 7. Pre-development (top) and contemporary proportional distribution of aquatic area.

The UMRS geomorphic reaches neatly superimpose on our floodplain inundation simulation (Figure 8). The Minnesota (XVI) and Chippewa River (XIV) Reaches show diverse inundation patterns, with the influence of the Chippewa River delta diminishing about mid-reach. The Wisconsin River Reach (XIII) is dominated by frequent floods, but the geomorphically diverse Maquoketa River Reach (XII) influences a diverse floodplain hydrology. The importance of the 2-year flood increases through the Iowa River (X), Des Moines River (VIII), and Quincy (VII) and Sny (VI) Anabranch Reaches until it meets the massive alluvial fan deposited by the Missouri River at Columbia Bottoms (V). Hydrology is similar to upstream reaches in the Jefferson Barracks Reach (IV) between the Missouri River and the Kaskaskia River (III) where the low elevation floodplain is greatly influenced by the 2-year flood. The Illinois River shows a relatively diverse flood stage distribution that is consistent in most of the reach (Figure 8). The influence of the higher head dams above river mile 150 is apparent, whereas the influence of dams is much less in most of the rest of the river. Dam effects on the Illinois River are exhibited by much larger and permanent backwater lakes compared to isolated lake and channel networks present at low flow prior to development (Mills et al., 1966).

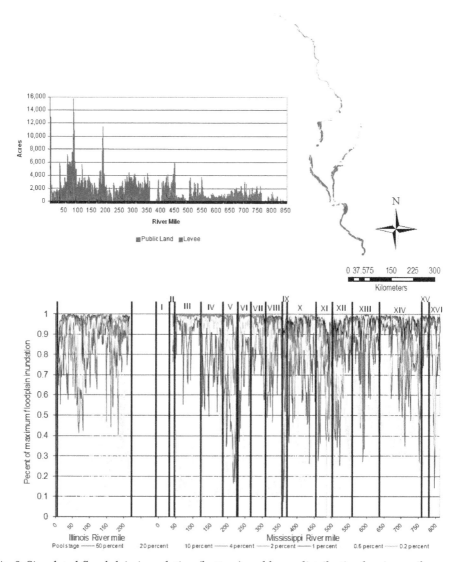

Fig. 8. Simulated floodplain inundation (bottom) and levee distribution by river mile.

Levees impede the flooding simulated above and prevent floodwater distribution in the floodplain south of river mile 450 (Figure 8). Most UMRS levee districts were established more than 100 years ago, and they occur as independent, quasi-political entities that have taxation and other authority for residents within their boundaries (Thompson., 2002). They have been hugely successful in preventing inundation during high frequency flood events with only a few significant disasters (Belt, 1975; Interagency Floodplain Management Review Committee, 1993; Galloway 2008). Levees and the development they protect have greatly altered hydro-ecological drivers and land cover in the floodplain.

3.3 Hydrogeomorphic methodology

Our HGM maps are relatively simple deterministic models that select various combinations of hydrology, geomorphology, and soil to map individual community distribution (Figure 9) which are integrated to produce potential vegetation estimates (Figure 9). Potential vegetation (HGM) maps (Figure 10) have been produced for several Mississippi River Reaches (Heitmeyer, 2008a; 2010) and many individual refuges or restoration sites

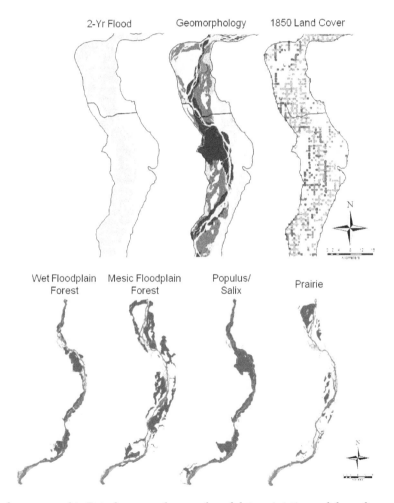

Fig. 9. Hydrogeomorphic Data layers and examples of deterministic model results.

(Heitmeyer and Westphal, 2007; Heitmeyer, 2008b). Each HGM evaluation is much more than simply combining GIS layers. An HGM evaluation reviews the physical setting, climate and hydrology, and the distribution and characteristics of presettlement habitats to establish a potential natural landscape. The HGM then reviews changes due to development and succession to make restoration and management decisions based on the likelihood of natural communities to recover from disturbance and in light of future disturbances. Potential vegetation maps assembled from hydrologic, geomorphic, and soils data are simply tools to visualize and quantify landscape response to management actions.

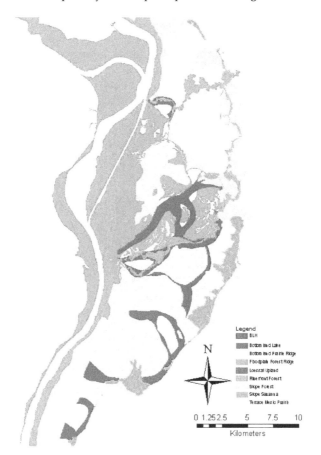

Fig. 10. A portion of a HGM map for the St. Louis region.

The near term intent is to complete an initial set of potential natural vegetation maps to help inform forest and land management plans for the entire UMRS (National Great Rivers Research and Education Center, 2010). The hydrology and geomorphology base layers described above were an important precursor to the rapid completion of the project. When the initial potential vegetation maps are complete, or as project needs dictate, potential vegetation maps for alternative floodplain management plans can be modeled to estimate environmental benefits that may accrue from restoration and management actions. Ultimately, these plant community models may be used in more comprehensive ecosystem services models that incorporate dynamic hydrology and ecosystem feedback loops that simulate complex functional processes of riverscapes (Thorp et al., 2006, 2008).

4. Discussion

There are many environmental and economic management needs that can be addressed with ecosystem modeling. Hydraulic models have become so precise that their results are routinely used for engineering design to simulate alternative design features (Silberstein, 2006). We believe the HGM approach for potential vegetation community assessment can achieve a similar standard for ecosystem restoration alternative analysis. The methods are not precise to species levels, nor very small spatial scale, at this stage of development but they do match well with the scale of most wildlife refuges and management areas that are the focus of most natural resource management and restoration activity. They also scale nicely for landscape ecology metrics and regional ecosystem management (USACE, 2011). HGM models have been developed for many floodplain systems (Klimas et al. 2009; U.S. Army Corps of Engineers, 2010), and they gain wide agency acceptance when developed collaboratively between managers and scientists.

These HGM methods for the UMRS are still quite simple in their statistical capacity and ability to model land cover occurrence. Future work will explore more rigorous landscape metrics that examine adjacency of land cover classes and associations with physical landscape features. The fundamental premise of the Hydrogeomorphic Method (HGM) is that vegetative communities segregate according to a single, or some combination of landscape features (e.g. geomorphology, hydrology, soil type). Indeed floodplain topography influences the frequency and duration flooding, which both directly influences plants via control over the length of oxic and anoxic phases, and indirectly influences plant communities by changing the physical properties of the soil (e.g. texture, pH, fertility). However, few studies have quantified the degree to which different plant communities segregate along key environmental gradients. By quantifying nonrandom associations among hydrology, soils and vegetation, land managers can increase their odds of successfully matching species and community types to suitable site conditions, thereby improving the odds of successful restoration.

To test the hypothesis that various plant communities segregate according to a given landscape feature or some combination of landscape features, an electivity index can be used (Jacobs, 1974; Jenkins, 1979; Pastor and Broschart, 1990). An electivity index calculates the juxtaposition of one cover type from one GIS data layer with some other landscape feature in a separate data layer.

These methods allow one to empirically test the hypothesis that a particular vegetation cover class 'elects' for a given landscape feature. If a particular cover class indeed elects for a given landscape feature, then it provides land managers with a prescription of broad-scale conditions that may be required for successful establishment of a given plant community under a given set of environmental conditions (Dr. Nathan DeJager, U.S. Geological Survey, Upper Midwest Environmental Sciences Center, La Crosse, Wisconsin, contributed text).

A multiple reference condition analysis has been proposed for UMRS ecosystem restoration planning (Nestler and Theiling, 2010). Sufficient data exist to evaluate hydrologic and geomorphic ecosystem drivers and land cover in presettlement, several historic snapshots, and contemporary conditions for nearly the entire 2.8 million acres. The virtual reference condition (i.e., simulated hydrology, potential vegetation, or geomorphic features), or plausible alternative future condition, is an important tool to estimate future without project condition and the response to alternative restoration plans (Figure 11; USACE, 2000). It is possible to simulate alternative floodplain management scenarios and extrapolate benefits as simple acreage estimates (Figure 11, bottom), potential vegetation (Heitmeyer, 2008; 2010), or any range of habitat suitability (USFWS, 1980) or ecosystem services metrics that can be attributed to potential land cover estimates.

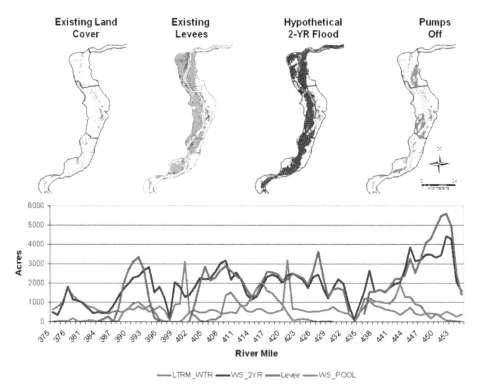

Fig. 11. Examples for UMRS benefits that may be attained by alternative floodplain management plans. LTRM_WTR = low flow surface water, WS_2YR = 50 percent exceedence/2-year flood, Levee = leveed area, WS_Pool = potential inundation under Pumps Off scenario.

5. Acknowledgements

An Army Corps of Engineers, Long Term Training Program grant from the Rock Island District and Headquarters supported much of Dr. Theiling's effort. The U.S. Army Corps of Engineers, St. Paul, Rock Island, and St. Louis Districts collaborated on archaeological investigations, flow frequency analyses, and flood risk assessments that provided data that were adapted for this study. The Upper Mississippi River Environmental Management Program administered by the U.S. Army Corps of Engineers and U.S. Geological Survey supported land cover mapping, future landscape analysis recommendations, and technical assistance. The U.S. Army Corps of Engineers, Navigation and Ecosystem Sustainability Program provided support to Dr. Theiling and Dr. Heitmeyer. Finally, The Nature Conservancy Great Rivers Partnership graciously supported development of the system-wide presettlement land cover data through a grant to the University of Wisconsin Department of Forest Ecology and Management.

Many individuals supported data development and analysis: John Burant – U.S. Army Corps of Engineers, Nathan DeJager – U.S. Geological Survey, Tim Fox – U.S. Geological Survey, Edwin Hajic – Illinois State Museum, Ken Lubinski – U.S. Geological Survey, David Mladenoff – University of Wisconsin - Madison, J.C. Nelson – U.S. Geological Survey, John Nelson – Illinois Nature Preserves Commission, Jim Ross – U.S. Army Corps of Engineers, Michael Reuter – The Nature Conservancy, Ted Sickley – University of Wisconsin - Madison, Paul West – The Nature Conservancy.

6. References

Barko, J., Johnson, B., & Theiling, C. 2006. *Environmental science panel report: implementing adaptive management.* U.S. Army Engineer Districts: Rock Island, Rock Island, IL; St. Louis, St. Louis, MO, and St. Paul, St Paul, MN.

Belt, C.B. 1975. The 1973 flood and man's constriction of the Mississippi River. *Science* 189:681–684.

Benda, L., Poff, N.L., Miller, D., Dunne, T., Reeves, G., Press, G., & Pollock, M. 2004. The network dynamics hypothesis: how channel networks structure riverine habitats. *BioScience* 54:413-427.

Bettis, E.A. & Mandel, R.D. 2002. The effects of temporal and spatial patterns of Holocene erosion and alluviation on the archaeological record of the Central and Eastern Great Plains, USA. *Geoarchaeology* 17:141-154.

Bettis, E.A., Anderson, J.D., & Oliver, J.S. 1996. *Land Sediment Assemblage (LSA) Units in the Upper Mississippi River Valley, United States Army Corps of Engineers, Rock Island District, Volume 1.* Quaternary Studies Program, Illinois State Museum, Springfield, Illinois. 39pp. + maps.

Bettis, E.A., Benn, D.W., & Hajic, E.R. 2008. Landscape evolution, alluvial architecture, environmental history, and the archeological record for the Upper Mississippi River Valley. *Geomorphology* 101:362-377.

Brinson, M. M. 1993. *A hydrogeomorphic classification for wetlands.* Army Engineer Waterways Experiment Station, Vicksburg, MS, USA. Technical Report WRP-DE-4.

Carlander, H.B. 1954. *A history of fish and fishing in the Upper Mississippi.* Special Publication of the Upper Mississippi River Conservation Committee, U.S. Fish and Wildlife Service, Rock Island Field Office, Rock Island, Illinois. 96 pp.

Chen, Y.H. & Simmons, D.B. 1986. Hydrology, hydraulics, and geomorphology of the Upper Mississippi River System. *Hydrobiologia* 136:5-20.

Christensen, N.L., Bartuska, A.M., Brown, J.H., Carpenter, S., D'Antonio, C., Francis, R., Franklin, J.F., MacMahon, J.A., Noss, R.F., Parsons, D.J., Peterson, C.H., Turner,

M.G., & Woodmansee, R.G. 1996. The Report of the Ecological Society of America Committee on the Scientific Basis for Ecosystem Management. *Ecological Applications* 6:665–691.

Clarke, S.J., Bruce-Burgess, L., & Wharton, G. 2003. Linking form and function: towards an eco-hydromorphic approach to sustainable river restoration. *Aquatic Conservation: Marine and Freshwater Systems* 13:439-450.

Council on Environmental Quality. 2009. *Proposed National Objectives, Principles and Standards for Water and Related Resources Implementation Studies.* The White House, Washington D.C.

Doyle, M.W., Stanley, E.H., Strayer, D.L., Jacobson, R.B., & Schmidt, J.C. 2005. Effective discharge analysis of ecological processes in streams. *Water Resources Research* 41: W11411, doi:10.1029/2005WR004222.

Fremling, C. R., Rasmussen, J. L., Sparks, R. E., Cobb, S. P., Bryan, C. F., & Claflin, T. O. 1989. Mississippi River fisheries: A case history. Canadian Special Publication of Fisheries and Aquatic Sciences 106.

Galloway, G.E. 2008. *A flood of warnings.* Washington Post. 25 June, 2008. ppA13.

Hajic, E.R. 2000. *Landform Sediment Assemblage (LSA) Units in the Illinois River Valley and the Lower Des Plaines River Valley.* Quaternary Studies Program, Technical Report 99-1255-16. Illinois State Museum, Springfield, Illinois.

Harwell, M. A., Meyers, V., Young, T., Bartuska, A. , Grassman, N., Gentile, J. H., Harwell, C. C., Appelbaum, S., Barko, J., Causey, B., Johnson, C., McLean, A. Somla, R., Tamplet, P., & Tosini, S. 1999. A framework for an ecosystem report card. *BioScience* 49:543-556.

Heitmeyer, M.E. 2008a. *An Evaluation of Ecosystem Restoration Options for the Ted Shanks Conservation Area.* Prepared for Missouri Department of Conservation, Jefferson City, Missouri by Greenbrier Wetland Services, Advance, Missouri.

Heitmeyer, M.E. 2008b. *An Evaluation of Ecosystem Restoration Options for the Middle Mississippi River Regional Corridor.* Greenbrier Wetland Services Report 08-02 for U.S. Army Corps of Engineers St. Louis District. Greenbrier Wetland Services, Advance, Missouri. 82pp.

Heitmeyer, M.E. 2010. *Feasibility Investigation: Hydrogeomorphic modeling and analyses Upper Mississippi River System Floodplain.* Prepared for U.S. Army Corps of Engineers, Rivers Project Office, West Alton, MO.

Heitmeyer, M.E. & Westphall, K. 2007. *An Evaluation of Ecosystem Restoration and Management Options for the Calhoun and Gilbert Divisions of Two Rivers National Wildlife Refuge.* Gaylord Memorial Laboratory Special Publication No. 13, University of Missouri-Columbia, Columbia, Missouri.

Interagency Floodplain Management Review Committee (IFMRC). 1994. *Floodplain Management into the 21st Century.* Report of the Interagency Floodplain Management Review Committee to the Administration Floodplain Management Task Force. U.S. Government Printing Office, Wishington, DC.

Jacobs, J. 1974. Quantitative measurement of food selection: a modification of the forage ratio and lvlev's electivity., index. *Oecologia* 14: 413-417.

Jenkins, S.H. 1979. Seasonal and year-to year differences in food selection by beavers. *Oecologia* 44:112-116.

Junk, W. J., Bayley, P. B., & Sparks, R. E. 1989. The flood pulse concept in river-floodplain systems. *Canadian Special Publication in Fisheries and Aquatic Sciences*. 106:110 – 127.

Klimas, C., Murray, E., Foti, T., Pagan, J., Williamson, M., & Langston, H. 2009. An ecosystem restoration model for the Mississippi Alluvial Valley based on geomorphology, soils, and hydrology. *Wetlands* 29:430-450.

Knox, J.C. 2007. The Mississippi River System. Pages 145 – 182 *in* A. Gupta (ed.) *Large Rivers: Geomorphology and Management*. John Wiley & Sons, Ltd., West Sussex, England.

Kondolf, G. M., Boulton, A. J., O'Daniel, S., Poole, G. C., Rahel, F. J., Stanley, E. H., Wohl, E., Bång, A., Carlstrom, J., Cristoni, C., Huber, H., Koljonen, S., Louhi, P. & Nakamura., K. 2006. Process-based ecological river restoration: visualizing three-dimensional connectivity and dynamic vectors to recover lost linkages. *Ecology and Society* 11(2): 5.

Laustrup, M. S., & Lowenberg, C. D. 1994. *Development of a Systemic Land Cover/Land Use Database for the Upper Mississippi River System Derived from Landsat Thematic Mapper satellite Data*. National Biological Survey, Environmental Management Technical Center, Onalaska, Wisconsin, May 1994. LTRMP 94-T001. 103 pp. (NTIS PB94-186889)

Leopold, L.B., Wolman, M.G, & Miller, J.P. 1964. *Fluvial Processes in Geomorphology*. Dover Publications, Mineola, New York. 522pp.

Lubinski, K.S. 1999. Floodplain river ecology and the concept of river ecological health. Chapter 2 *in* USGS ed. *Ecological Status and Trends of the Upper Mississippi River System 1998: A Report of the Long-Term Resource Monitoring Program*, Report Number LTRMP 99-T001, U.S. Geological Survey, Upper Midwest Environmental Sciences Center, La Crosse, Wisconsin. 236pp.

Lubinski, K. & Barko, J. 2003. *Upper Mississippi River – Illinois Waterway System navigation feasibility study: environmental science panel report*. U.S. Army Engineer Districts: Rock Island, Rock Island, IL; St. Louis, St. Louis, MO; and St. Paul, St Paul, Minnesota.

Madigan, T. & Schirmer, R.C. 1998. *Geomorphological Mapping and Archaeological Sites of the Upper Mississippi River Valley, Navigation Pools 1 -10, Minneapolis, Minnesota to Guttenberg, Iowa*. Report of Investigation 522,Prepared for U.S. Army Corps of Engineers, St. Paul district by IMA Consulting, Inc, Minneapolis, Minnesota. 285pp. + appendices.

Mills, H. B., Starrett, W. C., & Bellrose, F. C. 1966. *Man's effect on the Fish and Wildlife of the Illinois River*. Biological Notes 57. Illinois Natural History Survey, Urbana, Illinois. 24 pp.

National Great Rivers Research and Education Center. 2010. *Upper Mississippi River Systemic Forest Management Plan*. Prepared for U.S. Army Corps of Engineers, St. Louis District, River Project Office, West Alton, Missouri.

Newbold, R.V., O'Neill, J.D., Elwood, J.W., & Van Winkle, W. 1982. Nutrient cycling in streams: Implications for nutrient limitation and invertebrate activity. *American Midland Naturalist* 120: 628-652.

Newson, M.D. 2006. 'Natural' rivers, 'hydrogromorphological quality" and river restoration: a challenging new agenda for applied fluvial geomorphology. *Earth Surface Processes and Landforms* 31:1606-1624.

Norris, T. 1997. Where did the villages go?: Steamboats, deforestation and archeological loss in the Mississippi Valley. Pages 73–89 *in* A. Hurley (ed.) *Common Fields: An environmental history of St. Louis*. Missouri Historical Society Press, St. Louis.

Palmer, M.A., Bernhardt, E.S., Allan, J. D., Lake, P.S., Alexander, G., Brooks, S., Carr, J., Clayton, S., Dahm, C. N., Follstad Shah, J., Galat, D. L., Loss, S. G., Goodwin, P., Hart, D.D., Hassett, B., Jenkinson, R., Kondolf, G.M., Lave, R., Meyer, J.L., O'Donnell, T.K., Pagano, L., & Sudduth, E. 2005. Standards for ecologically successful river restoration. *Journal of Applied Ecology* 42:208–217.

Pastor, J. & Broschart M. 1990. The spatial pattern of a northern conifer-hardwood landscape. *Landscape Ecology* 4: 55-68.

Poff, N. L., Allan, J. D., Bain, M. B., Karr, J. R., Prestegaard, K. L., Richter, B. D., Sparks, R. E., & Stromberg, J. C. 1997. The natural flow regime: a paradigm for river conservation and restoration. *BioScience* 47: 769 784.

Rohweder, J., Rogala, J.T., Johnson, B.L., Anderson, D., Clark, S., Chamberlin, F., and Runyon, K. 2008. *Application of wind fetch and wave models for habitat rehabilitation and enhancement projects*. U.S. Geological Survey Open-File Report 2008–1200, U.S. Geological Survey, La Crosse, Wisconsin. 43 p.

Schulte, L.A. & Mladenoff, D.J. 2001. The original US Public Land Survey records: Their use and limitations in reconstructing presettlement vegetation. *Journal of Forestry* 99:5-10.

Sickley, T. & Mladenoff, D.J. 2007. *Pre-Euroamerican Settlement Vegetation Database for The Upper Mississippi River Valley*. Department of Forest Ecology and Management, University of Wisconsin-Madison, Madison, Wisconsin Draft Report to Paul West, The Nature Conservancy, Great Rivers Center for Conservation & Learning, Peoria, Illinois.

Silberstein, R.P. 2006. Hydrological models are so good, do we still need data? *Environmental Modeling and Software* 21:1340-1352.

Society for Ecological Restoration International Science & Policy Working Group. 2004. *The SER International Primer on Ecological Restoration*. Society for Ecological Restoration International. www.ser.org.

Sparks, R.E. 1992. The Illinois River-floodplain ecosystem. Pages 412 – 432 *in Restoration of Aquatic Ecosystems*. National Research Council, Washington DC. 528pp.

Sparks, R.E., Nelson, J.C., & Yin, Y. 1998. Naturalization of the flood regime in regulated rivers. *Bioscience* 48:706-720.

Stallins, J.A. 2006. Geomorphology and ecology: unifying themes for complex systems in biogeomorphology. *Geomorphology* 77:207-216.

Starrett, W.C. 1972. Man and the Illinois River. Pages 131-170 *in* R.T. Oglesby, C.A. Carlson, and J.A. McCann (eds.) *River ecology and man*. Academic Press, University of Wisconsin. 465 pages.

Theiling, C.H. 2010. *Defining Ecosystem Restoration Potential Using a Multiple Reference Condition Approach: Upper Mississippi River System, USA*. Ph.D. Dissertation, University of Iowa, Iowa City, Iowa. http://ir.uiowa.edu/etd/605/

Theiling, C.H. & Nestler, J.M. 2010. River stage response to alteration of Upper Mississippi River channels, floodplains, and watersheds. *Hydrobiologia* 640:17-47.

Theiling, C.H., Korschgen, C., Dehaan, H., Fox, T., Rohweder, J., & Robinson, L. 2000. *Habitat Needs Assessment for the Upper Mississippi River System: Technical Report*. U.S.

Geological Survey, Upper Midwest Environmental Science Center, La Crosse, Wisconsin. Contract report prepared for the U.S. Army Corps of Engineers, St. Louis District, St. Louis, Missouri. 248 pp. + Appendices A to AA. www.umesc.usgs.gov/habitat_needs_assessment/emp_hna.html

Thompson, J. 2002. *Wetlands drainage, river modification, and sectoral conflict in the Lower Illinois Valley, 1890 – 1930*. Southern Illinois University, Carbondale, Illinois. 239pp.

Thoms, M.C. 2003. Floodplain-river ecosystems: lateral connections and the implications of human interference. *Geomorphology* 56:335-349.

Thorp, J.H. & Delong, M.D. 1994. The Riverine Productivity Model: An Heuristic View of Carbon Sources and Organic Processing in Large River Ecosystems. *Oikos* 70:305-308.

Thorp, J.H., Thoms, M.C., & Delong, M.D. 2006. The Riverine Ecosystem Synthesis: Biocomplexity in river systems across space and time. *River Research and Applications* 22:123-147.

Thorp, J.H., Thoms, M.C., & Delong, M.D. 2008. *The Riverine Ecosystem Synthesis: Towards Conceptual Cohesiveness in River Science*. Elsevier, Academic Press, London. 208pp.

Trowbridge, A.C. 1959. The Mississippi in glacial times. *The Palimpest*. The State Historical Society of Iowa, Iowa City, Iowa 40:257-289.

U. S. Army Corps of Engineers. 2000. *Planning Guidance Notebook*. Engineering Regulation 1105-2-100. U.S. Army Corps of Engineers, Headquarters, Washington, D.C.

U.S. Army Corps of Engineers. 2003. *Upper Mississippi River System Flow Frequency Study Hydrology and Hydraulics Appendix C, Mississippi River*. U.S. Army Corps of Engineers, Rock Island District, Rock Island, Illinois. 120pp.

U.S. Army Corps of Engineers. 2004a. *Upper Mississippi River-Illinois Waterway System Navigation Feasibility Study with Integrated Programmatic Environmental Impact Statement*. U.S. Army corps of Engineers, St. Paul, Rock Island, and St. Louis Districts, Rock Island, Illinois. 621 pp.

U.S. Army Corps of Engineers. 2004b. *Upper Mississippi River System Flow Frequency Study: Final Report*. U.S. Army Corps of Engineers, Rock Island District, Rock Island, Illinois. 40pp.

U.S. Army Corps of Engineers. 2004c. *Improving Fish Passage through Navigation Dams on the Upper Mississippi River System*. Upper Mississippi River – Illinois Waterway System Navigation Study Interim Report, USACE, St. Paul, MN, Rock Island, IL, St. Louis, MO

U.S. Army Corps of Engineers. 2006. *Upper Mississippi River Comprehensive Plan*. Draft for Public Review. U.S. Army Corps of Engineers, Rock Island, Minneapolis, and St. Paul Districts, Rock Island, Illinois. May 2006. 121 pp.

U.S. Army Corps of Engineers. 2010. *Hydrogeomorphic Approach to Assessing Wetlands Functions: Guidebooks*. U.S. Army Corps of Engineers, Engineer Research and Development Center, Vicksburg, Mississippi. http://el.erdc.usace.army.mil/wetlands/hgmhp.html

U.S. Army Corps of Engineers. 2011. *Upper Mississippi River System Ecosystem Restoration Objectives 2009*. U.S. Army Corps of Engineers, Rock Island District, Rock Island, Illinois. http://www2.mvr.usace.army.mil/UMRS/NESP/Documents/UMRS%20Ecosystem%20Restoration%20Objectives%202009%20Main%20Report%20(1-20-2011).pdf

U.S. Fish and Wildlife Service. 1980. *Habitat Evaluation Procedures Handbook: Habitat as a Basis for Environmental Assessment: 101 ESM*. Division of Ecological Services, U.S. Fish and Wildlife Service, Department of Interior, Washington, D.C. http://www.fws.gov/policy/ESMindex.html

U. S. Water Resources Council. 1983. *Economic and Environmental Principles and Guidelines for Water and Related Land Resources Implementation Studies*. U.S. Water Resources Council, The White House, Washington, D.C.

Vannote, R.L., Minshall, G.W. , Cummins, K.W., Sedell, J.R., & Cushing, C.E. 1980. The river continuum concept. *Canadian Journal of Fisheries and Aquatic Sciences* 37:130-137.

Walters, C. 1997. Challenges in adaptive management of riparian and coastal ecosystems. *Conservation Ecology* [online]1(2):1. Available from the Internet. URL: http://www.consecol.org/vol1/iss2/art1/

Ward, J.V. & Stanford. J.A. 1983. Serial Discontinuity Concept of Lotic Ecosystems. Pages 29 – 42 in T.D. Fontaine and S.M. Bartell eds. *Dynamics of Lotic Systems*, Ann Arbor Science, Ann Arbor, Michigan.

Ward, J.V., Tockner, K., & Schiemer, F. 1999. Biodiversity of floodplain river ecosystems: ecotones and connectivity. *Regulated Rivers: Research & Management* 15:125–139.

Welcomme, R. L. 1979. *Fisheries ecology of floodplain rivers*. Longman, London, United Kingdom. 325 pp.

WEST Consultants, Inc. 2000. *Upper Mississippi River and Illinois Waterway cumulative effects study, Volume 1 and Volume 2*. Environmental Report #40 for the Upper Mississippi River – Illinois Waterway System Navigation Study. U.S. Army Corps of Engineers, Rock Island District, Rock Island, Illinois.

Whited, D.C., Lorang, M.S., Harner, M.J., Hauer, F.R., Kimball, J.S., & Stanford, J.A. 2007. Climate, hydrologic disturbance, and succession: drivers of floodplain pattern. *Ecology* 88:940–953.

Wiley, M.J., Osborne, L.L., & Larimore, R.W. 1990. Longitudinal structure of an agricultural prairie river system and its relationship to current stream ecosystem theory. *Canadian Journal of Fisheries and Aquatic Sciences* 47:373-384

Williams, B. K., Szaro, R. C., and Shapiro, C. D. 2009. *Adaptive Management: The U.S. Department of the Interior Technical Guide*. Adaptive Management Working Group, U.S. Department of the Interior, Washington, DC.

8

Introduction to Anthropogenic Geomorphology

Dávid Lóránt
Károly Róbert College,
Hungary

1. Introduction

In the past few decades interest in the environment has reached a peak as popular opinion has become aware of the extent of the human impact on natural systems. A proliferation of degrees has followed this wave of 'environmentalism', their focus has been on natural areas and the damage caused by human impacts. **Environmental geomorphology** is a special interaction of humans with the geographical environment which includes not only the physical constituents of the Earth, but also the surface of the Earth, its landforms and in particular the processes which operate to change it through time. This geographical environment can be investigated from several aspects:

- in the biological (ecological) approach emphasis is put on the biotic factors of the environment or on the structure itself;
- in the geographical approach research concentrates on the abiotic factors and functions;
- the technological or planning trend focuses the analysis on the economical-technical background of impacts.

To distinguish between the first two trends and the related disciplines, the terms (bio)ecology and geoecology are in use. The two concepts differ in handling the role of abiogenic and biogenic factors. In the past decade there was an intention to define geoecology as the study of abiotic factors and of issues concerning the functioning of the physical environment, while landscape ecology investigates the biogenic factors and problems of spatial organisation, structure. The far-reaching developments in the past one or two decades made landscape ecology become a wide theorethical-practical field of research, so the adaptation of international research results and educational experience is inevitable here too. The emerging science of landscape ecology is a tool for such studies and will be the cradle for advanced studies in the future.

Since the 1970s in the research of the physical environment two, frequently intertwining trends are prominent. One of them investigates the changes in the natural environment induced by human economic intervention (which are often undesirable) along with their counter effects. The other aims at the quantitative and qualitative survey of the resources and potentials of the physical environment and the evaluation of also regionally varying geographical potentials. **Anthropogenic geomorphology** is a new approach and practice to investigate our physical environment, because in the eighties the more and more urgent demands from society against geography - ever more manifest due to the scientific-technical revolution - underlined the tasks to promote efficiently the rational utilization of natural resources and potentials, to achieve an environmental management satisfying social requirements and opportunities.

The demand for complex environmental research has grown, since this is the only way to determine the loadibility of nature and the consequence of loading, to maintain the stable equilibrium of landscape, to preserve and develop the quality of life, and to give a long-term prognosis for the purposeful exploitation of environmental resources and potentials. Applying new methods and theories, the geography of today attempts to elaborate concepts and methods primarily novel in attitude to match the complex problems. As most of the problems of environmental management are, by their essence, interconnected by causal relationships, the solutions are justified, to be sought in the framework where the complex interrelationships of the human environment can be revealed in an integrated manner. All these, of course, do not mean to give up the investigation into the individual components of the environment, but these should be coordinated by one or several programmes which guarantee the study of the inner unity and multifarious nature of environmental factors and the detection of their interactions and development trends. The resulting environmental models may provide a uniform framework for basic (theoretical) and practical purpose research. We are convinced that any of the partical factors can only be studied in entirety and successfully if its relationships are known in the environmental systems.

Researchers reviewing the geomorphological literature of the last 40 years will gain the impression that the perception of Man as a geomorphological agent is a fairly recent development. We deal with anthropogenic geomorphology and we think that in an integrative study of this type, mankind must be regarded directly as a geomorphological agent, for it has increasingly altered the conditions of denudation and aggradation of the Earth's surface.

2. Scope of anthropogenic geomorphology

'The scope of anthropogenic geomorphology does not only include the study of man-made landforms but also the investigation of man-induced surface changes, the prediction of corollaries of upset natural equilibria as well as the formulation of proposals in order to preclude harmful impacts (Figure 1). The above topics and tasks make anthropogenic geomorphology a discipline of applied character. Its achievements should also serve – in addition to promoting the implementation of socio-economic tasks – environmental protection and nature conservation' (Szabó et al., 2010).

Generally the following fields of anthropogenic geomorphology have been identified:

Mining. The processes involved and the resulting landforms are usually called *montanogenic.*

Industrial impact is reflected in *industrogenic* landforms.

Settlement (urban) expansion exerts a major influence on the landscape over ever increasing areas. The impacts are called *urbanogenic.*

Traffic also has rather characteristic impacts on the surface.

As the first civilizations developed highly advanced farming relied on rivers, *water management* (river channelization, drainage) occupies a special position in anthropogenic geomorphology.

Agriculture is another social activity causing changes on the surface. *Agrogenic* impacts also include transformation due to forestry.

Although *warfare* is not a productive activity but has long-established surface impacts.

In contrast, the impacts of *tourism and sports* activities are rather new fields of study in anthropogenic geomorphology (Szabó et al., 2010).

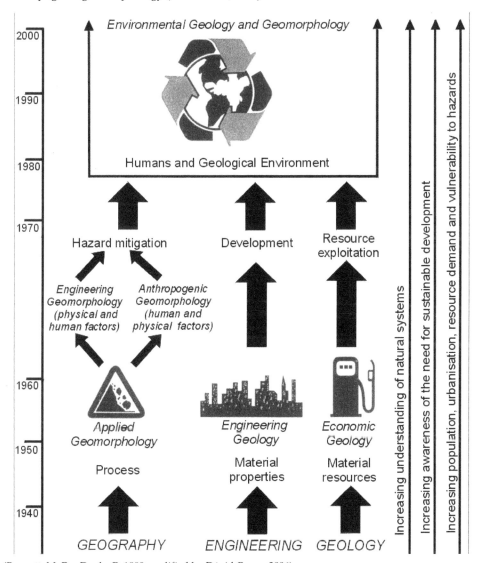

(Bennett, M. R. - Doyle, P. 1999 modified by Dávid-Baros, 2006)

Fig. 1. Development and differentiate of Earth Sciences (including Athropogenic geomorphology) and its connection with the environmental problems

3. Case study: Quarrying

3.1 Introduction to quarrying

This study intends to give an introduction to the significance of quarrying from the point of view of anthropogenic geomorphology, indicating the level of surface forming taken place due to the mining of these raw materials. The significance of this topic is supported by the presence of the so-called "mining landscapes" emerged since to the 19th century. The authors focus on the relevance of surface-forming by quarrying with special emphasis on factors influencing its spatial distribution, as well as on the characteristics and classification of mountainogenuous surface features of quarrying, giving an overview of the most important excavated and accumulated forms and form components on the macro, meso and micro levels. In the final section, some aspects of the opening and after-use of mining places are introduced by international and Hungarian case studies in order to observe how abandoned quarries can become assessed as „environmental values", and can be used as possible sites for exhibitions or for regional and tourism development projects.

The close relationship between mining activity and geology as well as geomorphology is not required to be explained in details, however, it should be mentioned that researchers only became interested in the problems of surface forming in a rather phase during the evolution of these sciences. It is well illustrated by *Figure 1* that scientific works on the research of landscape alterations caused by raw material production, can be traced back only from the 1960s to a greater extent both in the international and the Hungarian literature.

Mining activities were revolutionised by developments in mining techniques and the application of steam engines from the 19th century thus the exploitation of various mineral raw materials eventuated in the emergence of "mining landscapes". As a whole, the most generally mined raw materials for the building industry embrace the raw materials for the cement and lime industry, building and ornamental stones, sand and gravel as well as clay materials for the porcelain industry. This study intends to give an introduction to the significance of quarrying from the point of view of anthropogenic geomorphology, indicating the level of surface forming taken place due to the mining of these raw materials.

3.2 Surface-forming by quarrying

It can be claimed that the spatial distribution of quarrying, in general, is even in a sense that if geological conditions provided, settlements in which surroundings in any time of the history a quarry of any scale has not been opened can hardly be found in mountainous areas. When quarrying also aims to reach markets to a greater distance, instead of the factor mentioned above, market regulatories (economically exploitable supplies, transportation expenditure and possibility, etc.) become more important, thus in some cases, quarrying can show a rather high concentration in space. The level of socio-economic development being determinant for the quantity and quality indicators of the material flow between user and its environment, has undergone continuous changes during history. This is reflected by the extent of the mountainogenuous landforms on the one hand as well as in the rate of the expansion of areas effected by mining activity.

The site selection of quarries is, in addition to the geological conditions, also predominantly influenced by the topography of the area. Quarries are exploited by longwall face mining at mountainous or hilly terrains whereas at flat areas deep mining is applied, however, occasionally intermediate types can also be developed. Exceptionally, closed work is applied, too, as in the case of Fertőrákos (NW-Hungary). As far as quarries with longwall face mining are concerned, it is the topography that undergoes the most visible transitions as in some cases, face walls of several hundred meters in length and some ten meters in height can be resulted, even at more levels depending on the applied techniques of mining *(Figure 2)*.

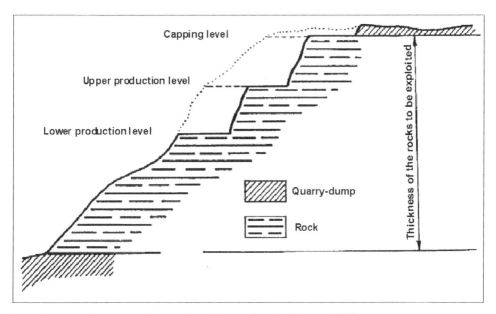

Fig. 2. Sitting of a quarry of several production levels (Ozorai, 1955)

In cases when the rock material to be exploited can be found under the ground level on a flat or on a declivate surface, we are constrained to open a quarry sunk under the surface. These types of quarries are sometimes created through the lowering of mine floor of quarries with longwall face quarrying. The occurrence of a very thick upper layer may necessitate the siting of the quarry actualised below the ground surface, thus in such cases exploitation takes place from underground shafts or rooms. Apart from this, the characteristics of the mined (metamorphic, igneous, sedimentary) rocks are of decisive relevance as well as the adherence of the various mine safety regulations. All of them may have an influence on the evolving forms too.

3.3 Characteristics and classification of the surface features of quarries

As a result of quarrying activity, it is the landscape morphology that undergoes the most visible changes *(Table 1)*.

Forms created as a result of mining can be classified into three main groups (*Dávid-Patrick 1998, Karancsi, 2000, Dávid, 2000*):

  a.) excavated (negative) forms,
  b.) accumulated (positive) forms,
  c.) and forms destroyed by quarrying activities can be classified into another group on the basis of the geotechnics of quarrying activities. This virtually means the levelling of the surface, which is called planation activity in geography.

FORM-MAKING ROLE OF QUARRYING ACTIVITIES			
A. On the basis of surface features			
EXCAVATED FORMS		**ACCUMULATED FORMS**	
on genetic basis and size			
EXCAVATED MACROFORMS (surfaces lacking in materials = caverns)		**ACCUMULATED MACROFORMS** (mine dumps) cone-shaped truncated cone-shaped terraced	
on the basis of mining techniques			
Simple excavated type: excavation pit delph	Complex excavated type: horizon mining	Simple accumulated type: single quarry dump	Complex accumulated type: quarry dumps in groups
EXCAVATED MESOFORMS mine wall debris apron mine floor		**ACCUMULATED MESOFORMS** plateau slope	
MICROFORMS			
Excavated microforms: rock buttress and pillar pinnacles rock benches small shallow ponds	microforms created as a result of natural processes: mass movements linear erosion	Accumulated microforms: heap boulder	
B. On the basis of the type of geotechnic activity			
PLANATION ACTIVITIES (PLANATION)			
ABRAIDING		**FILLING UP**	

(Dávid, 2000)

Table 1. Form-making role of quarrying activities

The morphological study of forms of quarrying was undertaken in three categories, distinguished on a genetic basis and size (*Figure 3*). It should be noted, however, that there are several different types of classification for these forms (*Erdősi, 1966, 1969, 1987, Karancsi, 2000*). An aspect may be for example the location of the quarry compared to geological formations and surface macroforms (*Erdősi, 1987*), as well as metric categories for the order of magnitude can also be created taking the characteristics of the given area into account (*Karancsi, 2000*).

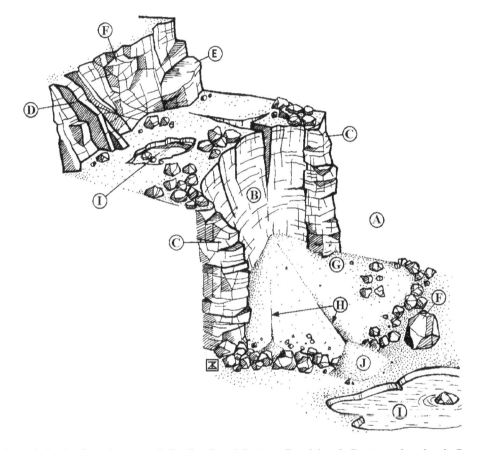

Legend: A-mine floor, B-mine wall, C-pillar, D-rock buttress, E-rock bench, F-out-weathered rock, G-talus slope, H-rainwater groove, I-depression with a small pond, J- debris cone

Fig. 3. Schematic layout of a quarry (Dávid-Karancsi 1999)

Macroforms are the most obvious landscape-forming remnants of mining having also an influence on landscape assessment. Excavated macroforms developed as a result of mining can virtually be regarded as surfaces lacking materials (caverns). Accumulated macroforms are the so-called mine dumps. Excavated macroforms created as a result of mining are composed of smaller elements (excavated mesoforms). Mine walls, mine floors and debris aprons can be distinguished in almost every mine. The morphological components of accumulated macroforms (the so-called mine and other dumps) are plateaux and slopes (accumulated mesoforms).

The surfaces of mesoform components can be divided into smaller and larger excavated depressions (possibly out-weathered sections) or accumulated elevations that are called microforms.

In addition to the influence of mining techniques and technology, and working rate, the characteristic features of the forms in all three categories are also determined by the geological characteristics of the area (structure, bedding), the feature of the rocks and the natural processes affecting them.

Excavated (negative) forms

The most common type of excavated form is an excavation pit or a delph in the surface (simple excavated type). Excavated macroforms of quarrying activities usually appeared before accumulated forms, therefore examples of them can be found in the first period of quarrying history. Mainly in the form of small quarries they can be found in the surroundings of almost every town and village situated in mountainous areas.

The other type of excavated forms is multi-levelled horizon mining (complex excavated type). The appearance of complex excavated forms is characteristic of modern times. The technical condition for the appearance of these forms was the increase in the capacity and efficiency of excavating equipment, and the geological conditions were the presence and exploration of thick stratums.

The form components of excavated mesoforms:

Mine wall: the steepest form component, whose angle of inclination with the mine flat is determined by the mining techniques (blasting, hand or power excavation) as well as the rock quality; normally it is nearly vertical. The mine floor is usually surrounded by mine walls on three sides.

Debris cones, debris aprons: a form component with a smaller angle of rest lying at the foot of mine walls whose material partly derives from mine working and partly from natural processes (rockfalls). They are initially developed by accumulation but as their origin, they are linked to excavation activities. With the amount of material in debris cones grown, they may coalesce to form a continuous debris apron.

Mine floor: an approximately flat ground surface surrounded by mine walls and debris aprons, in which form elements of a great variety (accumulations of the quarry material, quarry heaps, pillars, etc.) can be found.

The most common excavated microforms of mining are rock counterforts, rock benches, out-weathered quarry columns, pinnacles and pillars. These latter ones basically are the transitions between excavated and accumulated forms as being the positive remnant forms of quarrying. These forms can resist the damaging effects of natural processes we find only a little material derived from rock falls in front of them. In front of the wall sections next to them there are well-developed talus sloped originated in mass movements. Precipitation derived small shallow ponds can evolve in the holes of mine floor.

Accumulated (positive) forms

Accumulated macroforms are called quarry dumps. They are formed through the accumulation of materials currently of no value from an economic point of view. During outcast mining, dumps of various origins are heaped. By the burden-removal of layers covering the useful material, a significant amount of so-called sheating dump is created. This material (interstage and plant dump) can also be a result of the exploitation and processing of the material, i.e. during griding or crushing. The granulometric composition of quarry dumps is rather diverse, being influenced not only by geological conditions but also by the method of processing. There can also be different shapes of dumps, as curve-, fan- and round-shaped dumps created at the end of bankfills are distinguished in quarrying. In addition to these, temporary dumping of the quarry material also has to be mentioned as part of this group. They can be found singly (simple accumulated type) or in groups (complex accumulated type).

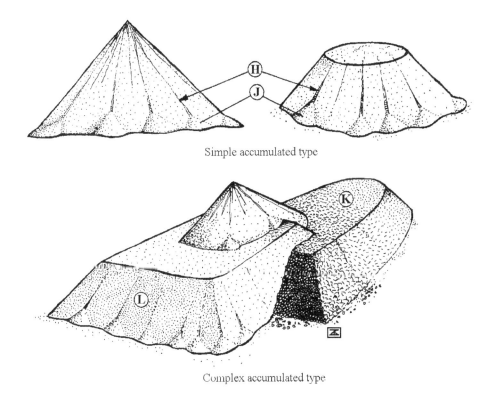

Simple accumulated type

Complex accumulated type

Legend: H-water groove, J- debris cone, K-plateau, L-slope

Fig. 4. Typical forms of quarry dumps (Dávid-Karancsi, 1999)

The shape of a positive form is determined by several factors: the original ground surface, the mode of accumulation and the physical features of the dump material. Cone-shaped, truncated cone-shaped and terraced dumps are the most common.

Form components of accumulated mesoforms

Plateau: the approximately flat ground surface surrounded by the slopes of dumps. Its extent is determined by the type of the dump. The largest plateaux can be found on terraced dumps. The plateaux of truncated cone-shaped dumps are usually smaller.

Slope: the sloping ground surface which surrounds the plateau or the peak in the case of a cone-shaped dump. Its angle of dip can vary within wide limits depending on the mode of accumulation, the dump material and the original ground surface.

The most obvious microforms of dumps, formed through natural processes, are rainwater grooves cut into slopes, which lie radially on cone-shaped or truncated cone-shaped dumps. The dump material carried by rainwater settles in small alluvial cones at the foot of slopes. Plateaux with approximately flat ground surface may be dissected by rainwater grooves cutting back into them.

The accumulated microforms of mine floors, formed as a result of mining activities, are the larger heaps and boulders cutting up the approximately flat ground surface.

Planation activities (planation)

Quarrying does not only have a form-making effect but it can also result in planation. With the spreading of dump material over natural or artificial dips (slopes, valleys, pits or depressions), they may be filled. Another possibility is the excavation of whole mountains during quarrying activities, resulting in a denudated surface. There are visible and advanced precedents of this in Hungary as seen on Photo 1.

3.4 Some aspects of the opening and after-use of mining places

Until recently, it has also been claimed that abandoned quarries both in Hungary or abroad were coupled by basically negative, unpleasant association of ideas, as antecedents has so far met mostly the "scars in the landscape" *(Photo 2)*.

However, an evolving new estimation can also be slowly observed, of which by abandoned quarries are regarded as „environmental values", assessed and used as possible sites for exhibitions or for regional and tourism development projects. The case studies below intend to give a short but detailed overview to two interesting examples.

Bluewater Shopping Centre

In recent years, many precedents, mainly from Great Britain show that commercial centres (hyper and supermarkets) are constructed in old quarries outside cities and towns, and connected to them facilities suitable for entertainment (parks, multiplex cinemas, gaming-rooms, concert halls, discos, galleries, art centres, etc.) are also developed *(Bennett-Doyle 1999)*. The most outstanding example for this is the Blue Water Shopping Centre located in Dartford at Junction No. 2. of the London Ring Road M25, marketed as the largest entertainment centre of this kind in Europe. This investment compelling both in its outside and inside appearance has been built between 1995 and 1999, in the area of the abandoned Blue Circle Chalk Quarry *(Photo 3)*.

Tokaj – Patkó Quarry

The quarry hosted an event of large scale on 30th June 2002 functioning as a „Festival cauldron". The event took place due to the fact that the Tokaj-Hegyalja Region was awarded World Heritage status in the category of cultural landscapes, thus programs were organised in the quarry strip pit to celebrate it *(Photo 4)*. The quarry since has been regularly used as a site for events.

Photo 1. Excavation of the Bélkő near Bélapátfalva (N-Hungary)

Photo 2. Scars of the andesite quarry at Sás-tó near Gyöngyös in the Mátra Hills (N-Hungary) made visible by clear-cutting with Mt. Kékes in the background

Photo 3. The Bluewater Shopping Centre near London with the mine wall of the Blue Circle Chalk Quarry in the background

Photo 4. Concert in the Patkó Quarry (Tokaj, N-Hungary)

4. Conclusion

Anthropogenic geomorphology is a new challenge for geomorphologists, since environmental problems have an effect on several branches of science. We teach anthropogenic geomorphology as an activity system, therefore, we believe in the equality ranks among the various fields of science in environmental protection and we assign an important part to anthropogenic geomorphology in the structure of our education.

5. Acknowledgement

This work has been supported by TÁMOP-4.2.1-09/1-2009-0001.

6. References

Bennett M. R.-Doyle P. 1999. *Environmental Geology: Geology and the Human Environment.* John Wiley and Sons, Chichester

Dávid, L. - Patrick, C. 1998. Quarrying as an anthropogenic geomorphological activity, In: *Anthropogenic aspects of geographical environment transformations* (Edited by József Szabó and Jerzy Wach), University of Silesia, Faculty of Earth Sciences,Sosnowiec, Kossuth Lajos University, Department of Physical Geography, Debrecen, Debrecen-Sosnowiec, pp. 31-39.

Dávid, L.-Karancsi, Z. 1999. Analysis of anthropogenic effects of quarries in a Hungarian basalt volcanic area, *2nd International Conference of PhD Students*, University of Miskolc, Miskolc, pp. 91-100.

Dávid, L. 2000. *A kőbányászat, mint felszínalakító tevékenység tájvédelmi, tájrendezési és területfejlesztési vonatkozásai Mátra-hegységi példák alapján*, PhD thesis, Debreceni Egyetem, Debrecen, 160. p. + supplements.

Erdősi, F. 1966. A bányászat felszínformáló jelentősége, *Földrajzi Közlemények* XIV., pp. 324-343.

Erdősi, F. 1969. Az antropogén geomorfológia mint új földrajzi tudományág, *Földrajzi Közlemények* XVII., pp. 11-26.

Erdősi, F. 1987. *A társadalom hatása a felszínre, a vizekre és az éghajlatra a Mecsek tágabb környezetében,* Akadémiai Kiadó, Budapest, 227p.

Karancsi, Z. 2000. A kőbányászat során kialakult felszínformák tipizálása. In: *A táj és az ember – geográfus szemmel* (Geográfus doktoranduszok IV. országos konferenciája, 1999. október 22-23.). Szeged, http://phd.ini.hu, CD-ROM

Ozorai, Gy. 1955. *A kőbányászat kézikönyve* I-II. Műszaki Könyvkiadó, Budapest, I: 423p. II: 399p.

Szabó, J.-Dávid, L.-Lóczy, D. (Eds.) 2010: *Anthropogenic Geomorphology: A Guide to Man-Made Landforms,* SPRINGER Science+Business Media B.V., Dordrecht-Heidelberg-London-New York

9

Anthropogenic Induced Geomorphological Change Along the Western Arabian Gulf Coast

Ronald A. Loughland[1], Khaled A. Al-Abdulkader[1],
Alexander Wyllie and Bruce O. Burwell
[1]Environmental Protection Department, Saudi Aramco
Saudi Arabia

1. Introduction

The term "Anthropocene" (the geological age since the industrial revolution) popularized by the Nobel Prize winner Paul Crutzen is apt for the change that has occurred along the Saudi Arabian Coast of the Arabian Gulf. Goudie and Viles (2010) expand on this concept emphasizing the impact of man on the landscape. This Chapter identifies and quantifies changes in the coastal ecosystems of the Arabian Gulf since 1967, and highlights those remaining valuable natural habitats and environmental assets. There is a paucity of data on coastal changes along the Western Arabian Gulf coast, possibly due to restricted access to coastal areas. The authors however have had a unique opportunity to undertake repeated visits to coastal areas and with the use of remote sensed imagery have been able to document changes.

The genesis of this coastline is principally due to the movement of the Arabian Tectonic Plate which began in the Middle to Late Cretaceous period when it collided with the Eurasian Plate. The result of the compression caused by this collision was the subduction of the Arabian Plate during the Tertiary and the formation of what is now termed the Zagros Fault Zone located in present day Iran (AlNaji, 2009, Haq and Al-Qahtani, 2005). This, with the separation of the African Continental Plate possibly due to the thinning of the crust (Seber et al., 2009, Hansen et al., 2006, Daradich et al., 2003) resulting in a north-westerly tilting and the formation of the Arabian Gulf basin. Erosion of the uplifted plate resulted in the deposition of paleo-deltaic features clearly visible in digital elevations models derived from the space Shuttle Radar Topography Mission (SRTM) data in the current day Eastern Province of Saudi Arabia. During the Quaternary the plate became more arid and the Aeolian influence became more dominant and resulted in the formation of massive ergs that are a major landform of the Arabian Peninsula such as the Rub Al Khali Desert. The west and east coasts of the Gulf are completely different in their structure and geomorphology because of their tectonic origins.

Coastal processes are largely wind driven as the maximum significant wave height is less than 1.5 meters; maximum mean wave period of 3-5 seconds; tidal speed less than 0.3 m/s; and maximum range in tidal water level is from 0.5 – 2.5 meters (Rakha et al., 2007) which are all slight when compared to open ocean levels. This low energy environment is confirmed by Summerfield (1991); and Viles and Spencer (1995). Current movements along the coastline are from north to south (Kampf and Sadrinasab, 2005, Brown, 1986). There are typical geomorphological long-shore drift features, such as the Ras Tanura Peninsula, as described by Davidson-Arnott (2010) that have been created by this anti-clockwise movement in the northern Arabian Gulf. These features are observed on the satellite imagery based maps of the northern coastline.

Parallel to the industrial growth of the region was the development of remote sensing technology that has allowed for a historic record of the impact of development on the coastal zone. This technology has provided a unique opportunity to document the changes in some coastal regions from a relatively pristine coastline to a highly developed coastal zone. The human population of the Saudi Gulf coastline has expanded exponentially, mostly as a result of hydrocarbon development in the Eastern Province. Residential, commercial, agricultural and industrial land use has thrived in a phase of economic development unknown in the history of Saudi Arabia and continued growth is predicted over the next two decades. The population in the Eastern Province grew from a few hundred thousand in the 1970s, approaching almost 3 million in 2010 (Lahmeyer, J.J., 2006). The Saudi Arabian Gulf coast has also changed significantly since 1967 as a result of the economic growth.

Development to meet increasing demands for residential, commercial and industrial expansion in the coastal zone is illustrated throughout the region. Recent changes in coastal geomorphology adjacent to highly developed coastal areas such as Dammam-Al-Khobar, Tarut Bay and Jubail are local examples of this development. Industry also places an enormous demand on the coastal zone for transport, ground water extraction, power generation, desalinization, sewerage and waste disposal.

Agriculture in the coastal zone (e.g., Qatif Oasis) has an indirect nutrient enrichment affect on the quality and productivity of marine waters such as those in Tarut Bay. The nutrient rich effluent originates from irrigation, particularly following fertilizer applications and can often enrich the natural biota within this coastal embayment (Basson, et. al. 1977). Map 1, Basic Land Cover and Land Use as at 2010, illustrates the development of the landscape in 2010 obtained from satellite imagery. The agriculture and development information in this map are digitized from enhanced satellite imagery; the sabkha derived by supervised classification techniques and density slicing; and partial development subjectively defined by infrastructure and land and water use.

This Chapter firstly describes the current situation and attempts to measure the change since the 1967 Corona imagery was acquired (Tarut Bay description is from 1934 and 1955). We then present a current risk matrix along with a short discussion on restoration programs, environmental awareness, public participation and the government legislative initiatives designed to protect the environment. The Chapter is concluded with a summary and conservation recommendations that will benefit the environmental assets of the Kingdom of Saudi Arabia.

2. Surface geomorphology of the Eastern Province of Saudi Arabia

The Eastern Province of the Saudi Arabian landform is characterized by sand movement across gravel plains, sabkha and limestone scarps. The landscape is dominated by sand sheets and widely rolling fossil sand dunes (Barth, 1999). The climate of the Eastern

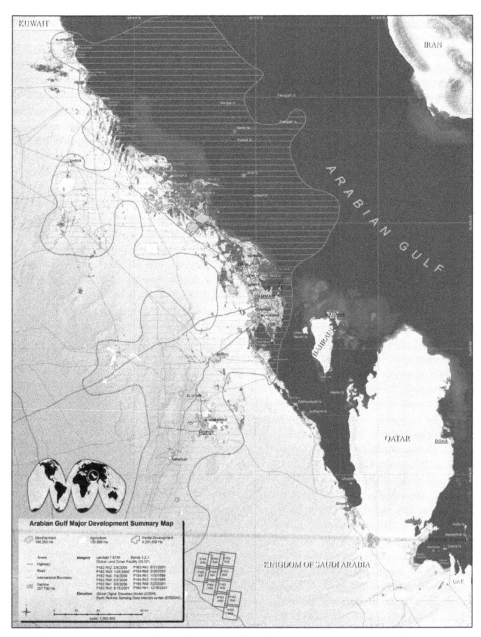

Map 1. Basic Land Cover and Land Use as at 2010. The background image of Map 1 is enhanced using a digital elevation model (DEM). Although the shadowing effect gives the impression of significant relief, in reality the landscape is relatively flat and has little impedance to the movement of Aeolian driven sands from the north to the south – see Section 2.

Province has changed over time with cyclic periods of changing temperature and moisture regimes (Chapman, 1971). It is currently in an arid phase where temperatures range from highs of 47°C in summer to lows of 3°C in winter with restrictive rainfall of less than 100 mm per annum. Wind patterns play a major role in the natural surface landscape development. Holm (1960), Chapman (1971) and Barth (1999 and 2001) all refer to the importance of the wind regime in the movement of sand across the region. The dominant winds come from high energy north-west direction sourced from the Mediterranean and occur from November to February with equally high energy summer Shamals from June to August and are primarily responsible for the inland movement of sand across the Province. Evidence of the directional influence of the prevailing winds creates striations caused by aeolian erosion and are clearly visible on exposed rock (see Figure 1). North of Dammam, coastal convection winds tend to move sand inward before it is transported south, whereas south of Dammam the shallower Gulf of Salwa has lesser convection influence and the sand dunes occur closer to the coast.

Fig. 1. Exposed Aeolian Weathering at Dhahran

Barth (2001) suggests four zones of sand balance within the Eastern Province (see Map 2). There is a potential sand deficit to the north; a negative sand budget in an area that is being denuded of vegetation caused by overgrazing resulting in the reactivation of fossil dunes; a balanced sand budget where sand coming in from the north is equal to sand leaving to the south; and a region of accumulation to the south. The background LANDSAT image also illustrates the generalized landforms based on the presence of absence of sand in the regime (Map 3). Another illustration of this sand movement across the region is the changes suggested by Barth (1999). He suggests that overgrazing has reduced the vegetation cover to such an extent that previously stable fossil dunes have once again become active. It is not unreasonable to assume that the dominant winds will likely increase the sand movement southwards – toward the heavily populated Jubail Industrial City and the Dammam metropolitan area. This assumption has health, logistic and maintenance repercussions on the populous of the region.

Map 2. Sand Transportation Zones affected by Prevailing Winds. (Adapted from Barth, 2001, Image Global Land Cover Landsat circa 2000)

Natural coastal change in this relatively low energy environment would involve the transportation of marine sediments and terrestrial sands from north to south over an extended period of time. Mobile dune systems would be expected to traverse the landscape and be initially arrested and slowed by sabkha and surface water with the following dunes continuing their southeasterly journey. However, transitional sands are being severely impacted through development requirements contributing to a lessening of the balanced sand budget referred to previously. The natural process pales into insignificance when the anthropogenic changes that have occurred since 1967 are taken into consideration.

Map 3. Generalized Geomorphologic Units of the Eastern Province (Image Global Land Cover Landsat circa 2000)

3. Development of the coastal zone

Some coastal habitats are under severe threat as a result of coastal development, and these areas are mostly within close proximity to human development or activities. Examples include salt marsh, mangrove and seagrass habitat areas. As mangroves have limited distribution, occurring in the central and northern areas within embayments surrounded by major development, they are under even greater risk of impact. The last major stands of mangrove occur in the Tarut and Musalmiyah Embayments with the largest trees occurring on and around Tarut Island. These mangrove ecosystems have been subjected to various impacts as a result of coastal development, with the most serious being smothering by land filling and as a result all mangrove habitats should be considered for immediate protection. Simultaneously aggressive mangrove rehabilitation programs need to be conducted to safeguard the genetic diversity of the mangrove populations and sustain fisheries and bird populations.

Subtidal muddy and sandy habitats are productive components of the marine ecosystem and are often ignored because attention usually focuses on more colorful and high profile habitats such as coral reefs. Seagrass habitats are widespread within the low energy subtidal areas of the coast and unfortunately vast areas of subtidal habitat are impacted each year through dredging and reclamation work. Reducing the footprint of dredging and reclamation operations on the productive shallow water habitats of the Gulf would be environmentally beneficial. Avoidance of nonessential marine dredging and reclamation works within these areas is the best strategy to minimize damage to these ecosystems.

4. Detecting change over time along the Gulf coastline

Aerial photography and freely available satellite imagery were used to determine coastal change. These types of data and related dates are listed in Table 1.

All LANDSAT satellite imagery used was downloaded with geo-location properties included. ASTER data were geo-referenced using the satellite ephemeral data and block adjusted where necessary. The 1934 aerial photography was 3rd order polynomially geo-referenced using common features located on the 1967 Corona imagery and the 1955 uncontrolled mosaic was triangulated to the same Corona data. All data were projected to UTM Zone 39 in WGS84 datum.

The coastline for each data set was determined using heads-up digitizing techniques from data enhanced to discriminate the land-water interface. Land use/cover classes (developed, agriculture and partial developed) were captured in the same way. Caution was exercised in compiling coastline changes due to the differing spatial resolutions of the imagery used; the methods of geo-location; and the state of the marine environment at acquisition. Some adjustments were necessary to increase the accuracy of the data. From the initial observations of the imagery it became obvious that there had been little change between 1934, 1955 and 1967 with minor changes detected in the 1972/73 data. A base coastline was established using the 1967 Corona data set that covered 95% of the coastal region of the Saudi Arabian Gulf. The area of sabkha was initially determined by Mah (2004) using supervised classification techniques. Small areas of in-fill were determined using a density slicing technique and both results combined into a single dataset. All coastal and land cover data were included in a database to calculate areas and distances, and where required statistics were calculated.

Date	Remote Sensing System	Spectral Attributes	Spatial Res	Comments
1934	Aerial photography	Panchromatic		Part of Tarut Bay
1955	Aerial photography	Panchromatic		Uncontrolled mosaic of Dammam Area and Tarut Bay
1967	Corona Imagery	Panchromatic	1 meter	95% Coastal Coverage
1972-1973	LANDSAT 1 & 2	Multispectral	60 meters	
1980	LANDSAT 3	Multispectral	60 meters	Sensor failing
1990	LANDSAT 4	Thematic Mapper	14.25 meters	USGS Mosaic
2000	LANDSAT 7	Thematic Mapper	14.25 meters	USGS Mosaic
2005	LANDSAT 7	Thematic Mapper	14.25 meters	Sensor failing
2005	ASTER	VISN	15 meters	Selected areas only
2006	ASTER	VISN	15 meters	Selected areas only
2009	GeoEye	Bands 1-3	0.6 meters	Selected areas only
2010	LANDSAT 7	Thematic Mapper	14.25 meters	Sensor failing

Table 1. Remote Sensing Systems Used to Determine Change

The Saudi coastline was divided into three coastal areas, Northern, Central and Southern coastlines. Data presented in this Chapter will illustrate changes in these coastal areas caused by development. A summary of changes along the entire Gulf coastline is also included.

5. Results

5.1 The Northern coastline

The upper part of the Northern Saudi coastline remains relatively undeveloped and contains interesting features such as mangrove and salt marsh habitats, coastal sand dunes, rocky shorelines, cliff formations and tidal creeks (Khores). Al Khafji Creek is the largest of the tidal creeks and has shoreward development. The mangrove and salt marsh habitats of Al Khafji Creek are threatened by this activity and other development, and the general discharge of waste. The remnants of an early mangrove plantation established in the 1980s appear to be growing well, however the need for protection against landfill encroachment at the head and mouth of the Creek is urgently required. There are around 2.5 ha of mangrove habitat in the Al Khafji tidal creek. Large expanses of salt marsh also occur, however, these continue to be impacted by ongoing land filling. This poses a serious risk to the Creek's ecosystem as there is a high probability that the Creek's waters and sediments will become polluted from leaching toxic materials sourced from illegally dumped wastes. Private sector partnerships are now undertaking a program to enhance the mangrove habitats of the Creek through annual mangrove plantations. It is envisaged that with the development of additional mangrove intertidal habitat, that the Creek's ecological value will increase, and the area will became an important nursery habitat for fish and crustaceans as well as migratory birds.

Tanajib – Manifa Embayment on the northern coastline is the most recently developed coastal area to facilitate the expansion of the oil industry. The embayment is very shallow and to allow oil operations in the area; and to minimize impacts associated with dredging, boat traffic, cable networks, pipelines and oil spills; a causeway and marine pads (small islands used for drilling) were developed (Map 4). To further reduce the impact of these structures on the marine ecosystem, the company responsible went to great lengths to minimize the overall footprint of the causeway and pads (width and length reduced by almost half), and there was no dredging allowed within the embayment. The direction and exact location of the causeway was also redesigned to minimize impact on subtidal habitats and the natural water circulation patterns. Water circulation was modeled and changes were made to the design of the causeway to maximize circulation, resulting in the causeway being built parallel to the water current direction, and incorporating 14 bridges. Existing fishing navigation channels remained intact, and constant monitoring was instigated during all construction and initial operation work to ensure compliance with marine environmental standards.

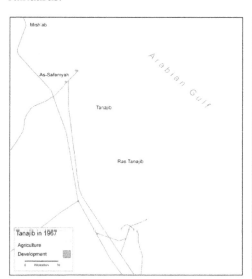 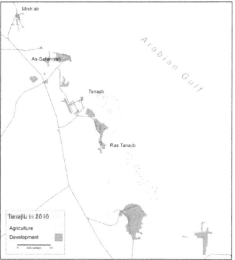

Map 4. Change in Land Use and Coastline for Tanajib from 1967 until 2010

The majority of the population in this area are employees who work at the Tanajib industrial facilities. The change in the district has occurred as the offshore oil reserves have required primarily industrial infrastructure development. This expansion included offshore causeways, rig pads and service facilities for production platforms. Development was estimated to cover only about 200 Ha in 1967 and apart from recent oil and gas processing facility construction of just under 12,000 Ha at Manifa, little has changed within the general coastal zone. Map 4 geographically illustrates these changes. Coastal infrastructure accounts for 1,100 Ha of environmentally sensitive causeways and drilling pads.

5.2 The Central coastline

This part of the coastline includes Jubail and Dammam areas, considered the most populated and developed areas of the Saudi coastline. Since the establishment of the Royal Commission in 1975, the Jubail area has experienced significant coastal development. The population of Jubail has grown from 140,800 in 1992 to 337,800 in 2010. It is expected that this increase will continue as the Industrial City is further developed. Unlike the Dammam area, there appears to be areas of the coastal zone relatively untouched. Map 5 illustrates the developed coastline and regions reclaimed from the marine environment to landfill and development. Much of these data are interpreted from satellite imagery from 1967 to 2010. From no development in 1967 there has been a steady growth to 44,000 Ha in 2010. There has been a small agricultural component of just over 2,500 Ha within this study area. A total of 4,100 Ha of coastal change has occurred. This development is set to continue with recent development announcements for the area. Map 6 clearly shows the development between 1967 and 2010.

Large areas of sabkha, salt marsh and mud flat remain relatively undeveloped to the north of Jubail Industrial City, although coastal recreational activity is apparent from field surveys. Reed (*Phragmitus australis*) growth occurs on the sewage outflows that spread over large areas of sabkha near to the coastline (see figure 2). This area is a haven for wildlife, and large numbers of birds of prey were observed hovering above its margins searching for food. The sewerage outflow, containing fresh water and nutrients, allows a wetland ecosystem to develop where it might not otherwise exist. Early in the development of Dhahran, it was recognized that the "Dhahran Ponds" were a source of food and a resting place for migratory birds (Evans, 1994) and were registered as a significant wetland.

Dredging at Jubail is significant, especially around Gurmah Island where there is a thriving mangrove community of approximately 63 ha. This mangrove community was severely polluted by oil as a result of the 1991 Gulf War, however it has since recovered. Map 7 illustrates the impact of dredging and reclamation in this region. Figure 3 shows the view from the developed shoreline east and north to Gurmah Island and the associated mangroves. The demarcation between the natural shallow water (light blue) and the dredged channel (darker blue) is apparent.

The embayment at Jubail has been largely unaffected by development except for dredging for landfill (Map 7). Mangrove, salt marsh and mudflat habitats are present in this area although they may not be as prolific as in Tarut Bay. There are a number of smaller areas within the Jubail region that are protected from the prevailing northerly winds. Seagrass, coral and salt marsh are all represented in these areas. The leeward side of Abu Ali Island protects 48 ha of mangroves from the incessant northern winds (shamal).

This entire embayment needs to be protected from further development, particularly any large industrial operations on the coastline. The Saudi Wildlife Commission has identified a large portion of this area (see Map 13 Section 6.3) as a proposed protected area. Gurmah Island with its rich mangrove habitats also needs protection, especially given that this island has significant mangrove research potential due to its polluted state during the 1991 Gulf war and its subsequent recovery.

Map 5. Developed Coastline at Jubail from 1967 until 2010 (Corona Imagery 1967)

Map 6. Change in Land Use and Coastline for Jubail from 1967 until 2010

Fig. 2. Reed Bed Sustained by Sewage Outflow Near Jubail.

Map 7. Dredging and Reclamation at Jubail Area (Image Global Land Cover Landsat circa 2000)

Fig. 3. View across the Dredged Channel to Gurmah Island and its Mangrove Habitat

5.3 Tarut Bay

The lower part of the Central Coastline is the Dammam area, which includes the highly productive Tarut Bay. The Saudi Wildlife Commission has identified Tarut Bay as a Resource Use Reserve. This is equivalent to the IUCNR classification for reserves as V – Protected Landscape or Seascape and VI – Managed Resource Reserve (Convention on Biodiversity, 2004; Sulayem and Joubert, 1994). The reserve category chosen is an attempt to prevent further deterioration of the Bay's environmental values while continuing to encourage its traditional exploitation. It does not seek to prevent development, but to have it occur with environmental values included in the development plan. It allows the traditional uses of the landscape or seascape to continue while discouraging massively destructive development.

The population growth of the Dammam area is illustrated in figure 3 with a rapid increase occurring from the 1970s. This growth was undoubtedly fuelled by the ongoing exploitation of oil and gas resources in the Gulf and subsequent significant global rises in oil prices with the Eastern Province being the center of this economic activity. Development has impacted the coastline at many locations as illustrated in Map 8. Landfill continues to expand into Tarut Bay and new canal projects are becoming a development trend in the Half Moon Bay area. Analysis of aerial photography and satellite imagery reveal that development in the Dammam region has increased from 5,000 Ha in 1967 to 70,000 Ha by 2010, while agriculture increased from about 8,500 Ha to 10,000 Ha mostly in the Tarut Bay area (however, the spatial context changes dramatically). Within the Tarut Bay area development has increased from 2,500 Ha to 28,000 Ha. There have been nearly 8,000 Ha of landfill expansion operations since 1955 for the Dammam area, while the Tarut Bay area totals 6,600 Ha. Each of the coastlines shows an ever increasing encroachment into the marine environment.

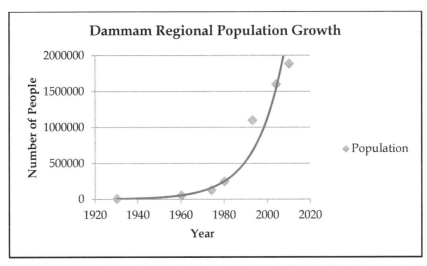

Fig. 3. Dammam Regional Population Growth from 1930 until 2010 (Data Source: Brinkhoff, 2011).

Tarut Bay is adjacent to the ancient Qatif Oasis, a natural fossil groundwater aquifer close to the surface that appears as a number of surface water pools, springs and small lakes. The development of the Oasis into a highly productive agricultural area to supply the needs of the growing population in the 1970 - 1990 period resulted in the use of chemical fertilizers, altered irrigation methods, and wetland drainage. Basson et. al. (1977) recognized that the high productivity of Tarut Bay was attributed to the influence of agricultural activity in the Qatif Oasis. The abundance of fish could be implied by the number of traditional fish traps (297) in the 1967 Corona satellite imagery. During this period there were still many hectares of healthy mangrove stands ideal as a protective nursery for many species of fish and crustaceans. It is estimated some 485 Ha (equivalent to 700 football fields (ff)) of mangrove have been lost and that 3,810 Ha (5450 football fields) of landfill development has encroached into the marine environment of Tarut Bay. Table 2 indicates the areal extent of the mangrove habitats at risk and the current ongoing threats.

The dramatic changes along the coast are clearly illustrated in Map 9a & 9b where land use and the actual coastline of 1967 have been extensively modified. The 2010 map shows the expansion of development. While some agriculture has been lost, the most significant modification has been to the natural coastline (Map 8). Map 10 shows successive landfill operations around Tarut Island. The result of such development is the loss of 134 Ha of mangrove forest (190 ff), an estimate of 440 Ha of salt marsh (630 ff) and 752 Ha of subtidal habitat (1075 ff) on Tarut Island alone.

What remnant mangrove habitats that remain in Tarut Bay are in urgent need of protection. In early 2011 the estimate of mangroves within this area was 325 Ha of varying density. This number included newly established mangrove plantation sites whose coverage is still very small at the present time. The mangrove habitat areas in immediate risk are estimated to be 89 Ha. These are at risk due to continued landfill and infrastructure development.

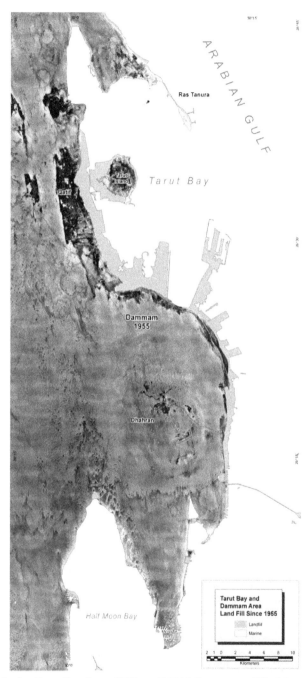

Map 8. Changes in the Coastline from 1955 until 2010 (Image modified from ARAMCO
Aerial Photography Mosaic 1955)

Location	Ha of Mangrove	Current Threats	Comments
Dammam Port	1 (1.5 football fields)	Road construction	Road widening for port traffic
Tarut Island	12 (17 football fields)	Landfill	Further expansion of landfill that will impact large areas of mangrove
Qatif	4 (6 football fields)	Landfill	Already closed from tidal flushing and remaining mangroves are under extreme stress
Safwa	72 (107 football fields)	Road construction	Direct landfilling and the resulting restricted water circulation following recent construction of a causeway (*See below)
Total	89 (131.5 football fields)		

This table is graphically presented in Map 12.
*If circulation within Raheema Bay is unaffected by the construction of a new causeway then the risk to mangroves decreases to 2 Ha.

Table 2. Mangrove at Risk in Tarut Bay

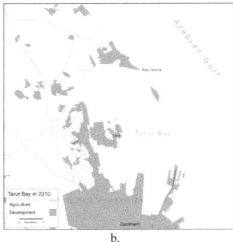

a. b.

Map 9. a. Development in Tarut Bay in 1967; b. Development in Tarut Bay in 2010

Map 11 illustrates a satellite image with the dredged marine environment overlaid. This dredging was initially undertaken for access to Port but became a source of material for the extensive landfill developments along the coastline over successive decades. Landfill material today is usually building rubble and other construction waste. Illegal dumping still occurs and a concerted effort is required to prevent this practice continuing. Map 11 also shows that only about 15% of the coastline of Tarut Bay remains undeveloped. Map 14 in

Section 6.3 illustrates the extent of a proposed reservation by the Saudi Wildlife Commission and the IUCNR (Saudi Wildlife Commission, 2010).

Within Tarut Bay there are very few undeveloped coastal landscapes. Maps 9a and 11 highlight these areas. This region is the major population center for the Eastern Province and as such has suffered the greatest anthropogenic induced change on the coastal environment. Map 12 illustrates the location and density of remnant, transplanted and new mangroves plantations within Tarut Bay. These maps also indicate those mangrove areas that are at immediate risk due to continuing impacts. There is a real possibility that the Raheema Bay mangroves could be seriously impacted if circulation is restricted by the new causeway presently under construction across the mouth of the Bay (Safwa – Raheema). For this reason all of the mangroves in Raheema Bay have been designated as at potential risk. Confidence is expressed in the environmental legislation and good environmental engineering design that should prevent such an occurrence. This prospective loss of mangrove habitat in Tarut Bay is indicated on Map 12 by a red outline to the icon representing the actual areal extent (Ha) of mangrove habitat at the different locations. It must also be noted that the self-propagation of mangroves indicates the resilience of the environment despite large scale developments. For example mangrove has shown resilience by self-sowing where none have existed before (e.g. Dammam Port). There are also concerted efforts within the Kingdom to recover and replace lost mangrove habitat. Design considerations for infrastructure in coastal areas are now also promoting environmental awareness and protection of habitats.

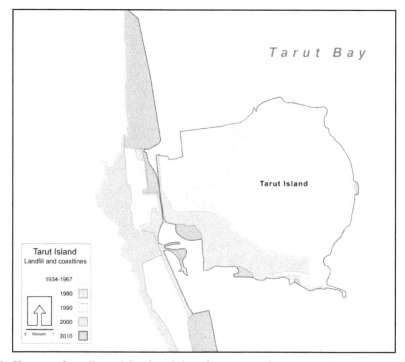

Map 10. Changes along Tarut Island and the adjacent coastline since 1934.

Map 11. Dredging Activities in Tarut Bay (Background image is Global Land Cover Landsat circa 2000)

Tarut Bay has been recognized as a landscape that requires protection, but still allows access for public use (IUCNR/MEPA, 1987).

Map 12. Mangrove Areas in Tarut Bay with Mangrove at Risk in Red

5.3 The Southern coastline

Unlike the northern and central coastlines, the southern coastline is mostly undeveloped and has less anthropogenic impacts. As such, the southern part of the Saudi coastline has been recommended as a nature reserve by the Saudi Wildlife Commission (Map 15). There has been some limited development at the northern end of this area such as at Qurayyah and Half Moon Bay consisting of salt water injection plants, power stations, desalination plants and recreation facilities. These facilities are relatively small in scale and scattered along the coast. They are required to support existing infrastructure within the central coastline areas around Dammam.

As a result of the areas relative pristine condition, wildlife extensively utilizes the coastal and marine areas of the southern region, with large bird populations inhabiting Zakhnuniyah and Judhaym Islands. These islands are of international significance as breeding sites for Socotra cormorant *(Phalacrocorax nigrogularis)*, and it is important that these islands be given protection from development or disturbance. The Gulf of Salwa is known as one of the most important sites in the world for the endangered dugong *(Dugong dugon)*, with this area containing significant seagrass habitat that enables the largest known single congregation of dugongs ever observed to exist in this area. Map 15 illustrates the reservation proposals from the Saudi Wildlife Commission for the Eastern Province including this important part of the Gulf coastline.

Al Uqair is a unique coastal port occurring on the coast of the Gulf of Salwa. The area has significant heritage sites of regional importance, consisting of an old trading fort and customs house area. This Port was once the major trading point from the Eastern Province to the then known world. In contrast to the high salinity of the Gulf of Salwa, the area contains fresh water springs with associated freshwater vegetation and fish inhabiting these areas. The area is also surrounded by sand dunes and has been recently proposed as a major tourist development site by the Saudi Commission for Tourism and Antiquities. In keeping with the Commission's philosophy the area's cultural and natural heritage should also be conserved as part of this development.

Ras Abu Qameess is the most isolated Gulf coastline of the Kingdom located between the United Arab Emirates and Qatar borders along the southern Gulf coastline. This rocky headland is in pristine natural condition and contains rich intertidal coastal rocky habitats, a large breeding population of Osprey *(Pandion haliaetus)* and Sooty Falcons *(Falco concolor)* and significant numbers of dugong *(Dugong dugon)*. There has been little development along this coastline. There are no major coastal encroachments and it has been recommended by the Saudi Wildlife Commission that a conservation reserve be established to protect this unique coastline in its natural condition (see Map 15 in Section 6.3).

6. Remediation programs

6.1 Coastal habitat restoration programs

Mangrove habitats have been the main focus for coastal restoration programs along the Gulf coast because they are an important intertidal habitat that is under threat from land development. It is estimated that 90% of the original mangrove ecosystems along the Gulf coast have been lost mostly as a result of coastal urban development. The current estimate of mangrove habitat within the Saudi Arabian territories of the Gulf is 412 ha, comprising 55 ha of plantation. Over the last few decades, influential private sector companies in association with partners such as the Ministry of Agriculture, King Fahd University of Petroleum and Minerals, Saudi Commission for Wildlife Conservation and local schools have been establishing mangrove plantations in the Eastern Province. The growth in some plantations has been promising and has resulted in private sector companies embarking on comprehensive mangrove restoration programs that include establishing strategic mangrove nurseries and plantation sites along the Eastern Province coastline. Restoration of mangroves needs to be undertaken on a large coordinated scale over an extended period and could also be supported by other industry and business in the region. Without this work there is a distinct possibility that mangroves would become locally extinct due to continuing development pressure.

6.2 Environmental awareness and public participation

The concept of public involvement with the management and conservation of the environment has been promoted as a potential solution to some environmental problems. Current public anti-littering and recycling campaigns, mangrove plantation programs and coastal clean-up events are examples of how the public can participate in protecting their environment. Environmental awareness requires a consistent and prolonged program of environmental education through the public schooling system. These topics need to be incorporated into the curriculum with teachers being trained (train the trainer) and students being assessed on their knowledge regarding their environment and their responsibilities towards protecting it. The goal should be on changing current behavior patterns that are not environmentally friendly. These same programs need to have a public focus, and the existing media sector is capable of achieving results that have proven successful for commercial marketing, by using national sport and entertainment personalities to promote good environmental behavior. As the majority of the Saudi population is under 25 years of age, these programs could very quickly change the environmental awareness within the Kingdom, with the same approach having proved successful in other countries. Public recognition of individual and community environmental achievements is also important to reinforce environmental responsibility, and government departments, educational institutions and private companies should all be encouraged to promote environmental champion programs among their respective staff and students.

6.3 Reservation proposals

Maps 13, 14 and 15 indicate the areas along the Arabian Gulf coastline that have been recommended for conservation. Some areas in the north and central regions such as Jubail and Tarut Bay have been nominated for multiple use reservations where sustainable utilization of the natural resources can continue along with the protection of the environmental, cultural and economic values of these areas. This is particularly true for the mainland coastal areas where there has been a long history of human settlement and natural resource exploitation. The offshore islands in the northern region are however designated due to their biodiversity value, with these areas being major centres of marine biodiversity in the region. Reservations in the southern region such as those proposed for the Gulf of Salwa and Ras Abu Qameess are nominated because these areas represent the last large areas where natural habitats dominate and wildlife populations flourish. The establishment of these reservations is required urgently if these areas's current values are to be protected indefinately.

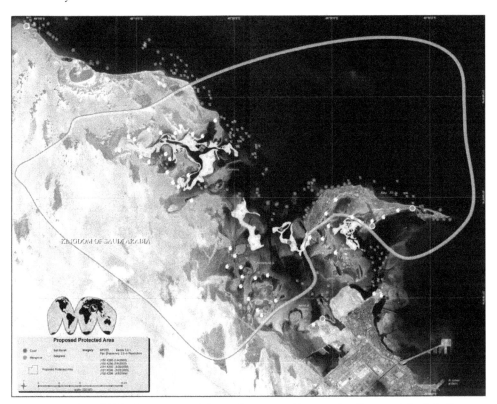

Map 13. Saudi Wildlife Commission Marine Sanctuary Reservation Proposal for Jubail.

6.4 Environmental legislation and law enforcement

There have been a number of government initiatives to protect the environment. The establishment of the Presidency for Meteorology and Environment (PME) has provided a guiding vehicle for legislation. Key documents such as the General Environmental Law, with rules for implementation and environmental protection standards provide guidance on protecting the environment (PME, 2004). The environmental law anticipates that the PME will liaise with other public authorities regarding the development and enforcement of environmental issues. The Saudi Wildlife Commission has created landmark documents such as the First Saudi Arabian National Report on the Convention of Biological Diversity. Saudi Arabia has also completed a National Biodiversity Strategy and Action Plan for the Country. There are 10 separate legislative acts dealing with the protection of the environment (AbuZinada et al. 2004). The Fishing Exploitation and Protection of Live Aquatic Resources in the Territorial Waters of Saudi Arabia Act – 1987, the Wildlife Protected Areas Act – 1995 and the Environmental Code – 2002 are relevant to the marine

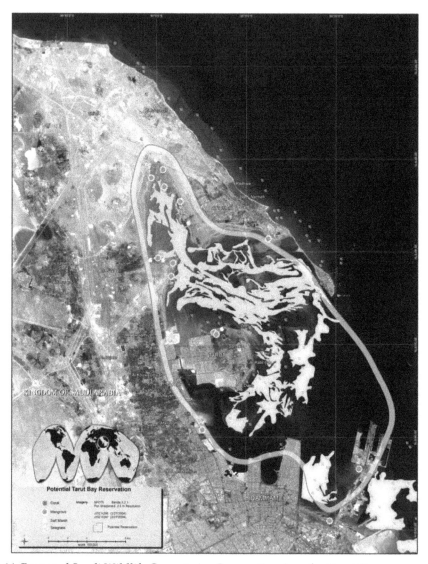

Map 14. Proposed Saudi Wildlife Commission Reservation Area for Tarut Bay

environment. There are prescribed prison sentences and fines for breaches of these acts (Evans and Abuleif, 2003). The National Coastal Zone Management Plan (NCZMP) is based on work undertaken by the PME and International Union for the Conservation of Natural Resources (IUCNR) in 1987 where Coastal Zone Management Programs were produced for the Red Sea and the Arabian Gulf (NCWCD, 2004). The plans were submitted for approval in 2003 (AbuZinada et al. 2004). The Gulf coastline has some of the most unique and productive habitats in the region and the Kingdom has appropriate legislation in place to protect it including a coastal setback policy of 400m (Royal Decree 1004 20/1/1419H) for any new developments, and an approval process (Committee of Four - Royal Decree 982/M 15/9/1419H) for any projects (particularly land filling) in the coastal zone. This Committee of Four consists of the Ministry of Municipal and Rural Affairs, Coast Guard, PME, and Ministry of Agriculture.

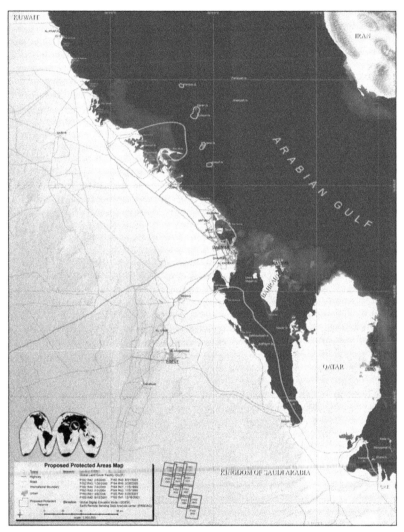

Map 15. Proposed Saudi Wildlife Commission Conservation Reservations for the Eastern Province

7. Conclusions and conservation recommendations

This Chapter reports the rapid change from 1967 for the Arabian Gulf coastline of Saudi Arabia and highlights the development trends and land use changes until 2010 within the coastal zone. This Chapter indicates areas where urgent additional conservation actions are now required to protect the remaining natural habitats and wildlife populations from continued impact resulting from rapid and hastening coastal development.

Serious consideration should be given to prevent coastal development for industrial purposes, and instead have this development moved inland with resources currently used for coastal landfill employed instead to create inland connective waterways. This would free up these productive and attractive coastal regions for conservation, recreational and managed tourism and residential land use. This would significantly enhance the quality of life for the residents of the Eastern Province.

Other areas of the coastline are also extremely important for wildlife, and these areas tend to be the more remote locations such as the Offshore islands and isolated coastal regions such as the Gulf of Salwa and Ras Abu Qamees. Wildlife populations flourish in these areas due to the lack of physical disturbance from human related activities and also due to the large areas of relatively undisturbed habitats. These remote nesting, roosting and refuge locations, coupled with an abundance of nearby foraging resources results in very large populations of marine mammal (dugong and dolphin), marine turtles and marine birds inhabiting and breeding each year in these areas. These populations are not only significant within the Kingdom, but are also of regional and international importance. With increasing human populations and increased development pressure on the coastal zone, these remote undisturbed areas are becoming smaller and geographically closer to human settlements, and eventually wildlife populations will have no remaining undisturbed wilderness areas in which to exist, and this will be a great loss for the Kingdom, particularly for its future generations.

The Kingdom of Saudi Arabia has good environmental strategies and legislation in place to protect the environment and these strategies and legislation only need to be effectively enforced. Changing public attitudes and behavior towards the environment is also a major objective that needs immediate attention with the development and implementation of environmental awareness and education programs aimed at all levels of society.

There are existing proposed conservation reservation areas along the Gulf coast (Maps 8 and 13) and these incorporate the remaining few natural regions of the Gulf coastal zone including Ras Abu Qamees, the Gulf of Salwa, Tarut Bay, the Jubail Embayment and the offshore coral islands (Map 15). These recommendations should be immediately reviewed and alterations made that ensure the conservation of these areas' biodiversity is guaranteed while at the same time promoting sustainable development and use by the areas' stakeholders. Conservation reservations do not have to exclude all other uses and activities, and in fact some hydrocarbon coastal reservation areas are testament to this, with these areas containing the greatest proportion of undisturbed coastline and intact habitats (e.g., mangroves) than many other areas along the Gulf coastline. Good environmental stewardship can allow nature conservation and sustainable development to flourish within the same geographical areas, it is really all about management of the area's natural resources to ensure that their inherent values are maintained and where possible enhanced. The earlier recommendations of the United Nations Environmental Program for Tarut Bay had this particular notion in mind when they nominated Tarut Bay as a Class V and VI reservation. This identified the Bay as a significant public use and amenity resource where traditional activities such as fishing could continue, and new developments that were in keeping with the Bay's environmental uniqueness and values could be undertaken, with the collective goal of protecting this productive Bay for present and future generations.

8. Acknowledgements

We acknowledge the contribution to this paper by Abdullah Mah who assisted in identifying and quantifying remnant mangroves in the region; provided data on sabkha; and contributed to the discussion. Saudi Aramco is also acknowledged for its Environmental Protection Program and the dedication of its staff to protect, restore and establish ecosystems vital for the ongoing health of the environment.

9. References

AbuZinada, A.H., Robinson, E.R., Nader, I.A. and Al-Wetaid, Y.I. (c2004) *First Saudi Arabian National Report on the Convention on Biological Diversity.* National Commission for Wildlife Conservation and Development, 131 pages.

AlNaji, Nassir (-) A Brief Tectonic History of the Arabian Basin, Master's Thesis Extract. http://strata.geol.sc.edu/Nassir-Thesis-SITE/CHAPT04.html accessed 22 April 2009.

Barth, Hans-Jörg (2001) Characteristics of the wind regime north of Jubail, Saudi Arabia, based on high resolution wind data. *Journal of Arid Environments,* 47:387-402.

Barth, Hans-Jörg (1999) Desertification in the Eastern Province of Saudi Arabia. *Journal of Arid Environments,* 43:399-410.

Basson, P.W., Burchard Jr, J.E., Hardy J.T. and Price, A.R.G. (1977) *Biotopes of the Western Arabian Gulf.* Dhahran, Saudi Arabia: Aramco Department of Loss Prevention and Environmental Affairs, 284p.

Brinkhoff, Thomas (2011) City Population Web site www.citypopulation.de Accessed April 2011.

Brown, Roger (1986) The Content and Nature of Arabian Gulf Seawater. *Bulletin 29, Emirates Natural History Group,* July 1986.
    http://www/enhg.org/bulletin/b29/29_05.htm

Chapman, Randolph W. (1971) Climate Change and the Evolution of Landforms in the Eastern Province of Saudi Arabia. *Geological Society of America Bulletin* v 82 p. 2713-2728.

Convention on Biological Diversity (2004). Thematic report on protected areas or areas where special measures need to be taken to conserve biological diversity. *Convention on Biological Diversity* website accessed October 2004.
    http://www.biodiv.org/doc/world/sa/sa-nr-pa-en.pdf

Davidson-Arnott, Robin (2010) *Introduction to Coastal Processes and Geomorphology,* University Press Cambridge, ISBN 978-0-521-87445-8 (Hb) ISBN 978-0-521-69671-5 (Pb), Cambridge, United Kingdom.

Daradic, Amy, Mitrovica, J. X., Pysklywec, R. N., Willett, S. D. and Forte, A. M. (2003) Mantle Flow, Dynamic Topography, and Rift-flank Uplift of Arabia. *Geology,* October 2003; v. 31; no. 10; p. 901-904.

Evans, M.I. (1994) *Important birds areas in the Middle East.* Cambridge: Birdlife International, 410p.

Evans, J.R. and Abuleif, K.M. (2003) New Kingdom environmental regulations on the way. In *EnviroNews* Issue No. 10 Summer 2003, Aramco Publication, p5.

Goudie, Andrew and Viles, Heather (2010) *Landscapes and Geomorphology – A Very Short Introduction,* Oxford University Press, ISBN978-0-19-956557-3, New York, USA.

Hansen, Samantha, Schwartz, Susan, Al-Amri, Abdullah and Rogers, Arthur. (2006) Combined Plate Motionand Density-driven Flow in the Asthenosphere Beneath

Saudi Arabia: Evidence from Shear-wave Splitting and Seismic Anisotropy. Geology, October 2006 V. 34 no. 10 p. 869-872; doi: 10.1130/G22713.1.

Haq, Bilal U. and Al-Qahtani, Abdul Motaleb (2005) Phanerozoic Cycles of Sea-level Change on the Arabian Platform. *GeoArabia*, Vol. 10, No.2 2005.

Holm, Donald A. (1960) Desert Geomorphology in the Arabian Peninsula. *Science, New Series*, Vol. 132, No. 3437, pp. 1369-1379.

IUCNR/MEPA (1987) *Saudi Arabia: An assessment of coastal zone management requirements for the Red Sea*, Arabian Gulf,*Saudi Arabia: An assessment of biotopes and coastal zone management requirements for the Arabian Gulf* [Table of contents only], *Saudi Arabia: An assessment of national coastal zone management requirements* [Table of contents only]. Technical reports No. 3, 5 and 7. Gland, Switzerland: International Union for Conservation and Natural Resources.

Kampf, J. and Sadrinasab, M. (2005) The Circulation of the Persian Gulf: A Numerical Study, *Ocean Science Discussions*. 2, pp129-164, May 2005.

Lahmeyer, J. J. (2006) Population Statistics Webpage Accessed February 2011.
    http://www.populstat.info/Asia/saudiarc.htm

Mah, A. (2004) Detecting mangrove changes using Landsat imagery: Tarut Bay Pilot Project. In *EnviroNews*, Issue No. 12, Saudi Aramco publication, p1-3.

NCWCD – National Commission for Wildlife Conservation and Development (2004) NCWCD and its mandate from *NCWCD* website accessed October 2004 http://www.ncwcd.gov.sa/english/intro.htm

PME – Presidency of Meteorology and Environment (2004) Environmental Protection Standards. *From Presidency of Meteorology and Environment* website accessed September 2004. http://www.pme.gov.sa/Env.asp

Rakha, K, Al-Salem, K. and Neelamani, S. (2007) Hydrodynamic Atlas for the Arabian Gulf. *Journal of Coastal Research*, Special Issue 50, pp 550-554, 2007.

Saudi Wildlife Commission (2010) Website accessed February 2011

Seber, Dogan, Vallve, Marisa, Sandvol, Eric Steer, David and Barazangi, Muawia. (1997) Middle East Tectonics: Applications of Geographic Information Systems (GIS). *Institute of the Study of the Continents*, Cornell University, Snee Hall, Ithaca, NY. http://atlas.geo.cornell.edu/htmls/gsa_today.html accessed 22 April 2009 http://www.swc.gov.sa/English/default.aspx

Sulayem, M. and Joubert, E. (1994) Management of protected areas in the kingdom of Saudi Arabia. *Unasylva* - No. 176 - Parks and protected areas Vol. 45 - 1994/1. Food and Agriculture Organization of the United Nations (FAO) http://www.fao.org/docrep/v2900E/v2900e08.htm

Summerfield, Michael A. (1991) *Global Geomorphology*. Pearson Education Limited, ISBN 0-582-30156-4, Harlow, Essex, England.

Viles, Heather and Spencer, Tom (1995) *Coastal Problems – Geomorphology, Ecology and Society at the Coast*. Edward Arnold, ISBN 0-340-53197-5 (Pb), ISBN 0-340-62540-6 (Hb), London, UK.

10

From Landscape Preservation to Landscape Governance: European Experiences with Sustainable Development of Protected Landscapes

Joks Janssen[1] and Luuk Knippenberg[2]
[1]Land Use Planning Group, Wageningen University, Wageningen,
[2]Centre for International Development Issues, Radboud University Nijmegen,
The Netherlands

1. Introduction

Over the last two decades there has been a significant increase in the appreciation of the cultural landscape by the public and by politicians; this phenomenon is taking place in most European countries. In 1992 the importance of cultural landscapes was recognized on an international scale with their inclusion in the World Heritage Convention. Eight years later, in 2000, the Council of Europe adopted a European Landscape Convention (ELC) and presented it to member states for adoption. Through innovations such as the World Heritage Convention and the European Landscape Convention, cultural landscape has become increasingly central to matters of sustainability and place-making across both urban and rural realms. As a consequence, the thinking on protected areas has undergone a fundamental shift. Cultural landscapes are at the interface of nature and culture. Therefore, both natural and cultural resource conservation converge, creating opportunities for collaboration.

In Europe, the approach to protecting landscapes has generally been one of 'designation', that is, drawing lines round areas valued by experts. The 'designation' approach, however, has come under criticism for a number of reasons, not least the growing realization that neither the ecologic and geomorphologic nor the axiological and aesthetic aspects of landscapes can be safeguarded in the long term on the basis of corralling stand-alone sites. Modern aesthetic, geomorphologic and ecologic objectives rely on a site-in-context approach based on a concern for visual, morphologic coherence and ecological connectivity across the wider countryside. Whereas protected areas were once planned against people, now it is recognised that they need to be planned with local people, and often for and by them as well. Instead of setting landscapes aside by 'designation', nature and landscape conservationists now look to develop linkages between strictly protected core areas and the areas around: economic links which benefit local people, and physical links, for instance via ecological corridors, to provide more space for species and natural processes. As a result, landscape conservation of continuously evolving landscapes is about the management of change – the landscape should not become frozen but kept alive (Bloemers et al., 2010).The people that live and work in landscapes, be it farmers, residents or entrepreneurs, have to be actively involved in formulating conservation plans, collective decision-making and the performance of landscape management measures.

It is now recognized that protected landscapes (IUCN Category V Protected Landscape/Seascape) and cultural landscapes share much common ground: both are focused on landscapes where human relationships with the biotic and abiotic natural environment over time define their essential character. They can help to conserve both wild biodiversity and agricultural biodiversity, and to conserve human (cultural) history alongside the geomorphologic past (Reynard, 2005; Panizza 2009; Farsani, Coelho and Costa, 2011). Against this background, protected landscapes throughout Western Europe more and more function as flagships for a new and integrated urban-rural public policy. Since landscape conservation and environmental government are aspects of a single whole, conservation, increasingly is seen as an integral part of sustainable management. This is highlighted by a range of protected landscapes in Western Europe, which, since the 1990s, strive towards a regional integration of agriculture, nature and landscape, thereby overcoming the often strong sectoral division of countryside and town planning and natural and cultural resource management.

The adopted approaches for protected landscapes in Europe increasingly recognise the critical links between nature, culture, and community for a long-term sustainable development. Landscape management plans and projects seek to support a 'virtuous circle' in which the socio-economy of a region contributes to nature and beauty, and the environment underpins community and prosperity of the protected landscape (Powell et all, 2000; Selman, 2006). Knowledge about the spatial-temporal aspects of the metabolism between nature and society is needed in order to support this 'virtuous circle'. It is precisely the hybrid character of landscape, that is, that societal and "natural" factors are intrinsically linked to one another that ensure that cultural, aesthetic, economic and social dimensions are as much involved as ecological functioning or abiotic, morphological conditions. Landscape, as a realm of this hybrid human-environmental interaction, is at the centre of sustainability and sustainable development (Wascher, 2000; Reynard and Panizza, 2005).

The re-positioning of cultural landscape within the sustainable development agenda is opening up new challenges for landscape governance. The term landscape governance reflects two contemporary, interrelated changes in the scale and organisation of decision-making about the landscape (Beunen & Opdam, 2011). Government power is decentralized to the lower tiers of command, while a growing number of private parties and citizens begin to actively participate in decision-making. As a result, the term governance has been introduced in the field of protected areas and the term 'protected area governance' has recently been established (Borrini-Feyerabend, 2004; Dearden & Bennett, 2005; Fürst et al. 2006, Stoll-Kleemann et. al., 2006). A cornerstone was the Vth IUCN World Parks Congress in Durban 2003; since then the topic of governance has also been applied to different categories of protected areas, including protected landscapes and, more recently, so-called Geoparks. However, a scientific discussion concerning governance in protected landscapes is still missing.

European landscape conservation is a practice in the making, continuously evolving because of changing political and institutional contexts, new insights in the dynamic relation of

society and nature, and the unfolding of new vectors for regulating socio-economic, cultural
and environmental change. In this chapter some European experiences with landscape
protection are described and analyzed. On the basis of English-language literature this
chapter examines the western European experience with landscape conservation as well as
the related governance issues, shifting from 'preservation by designation' to 'conservation
through development'. We trace the different conservation attitudes in Western Europe, as
well as the subsequent conservation systems that have been created for sustaining cultural
landscapes. We focus on Britain, France, and Germany, because of the long history of
preserved landscapes in these countries and the relatively large areas of protected
landscapes that are managed for recreational, scenic, educational, and heritage purposes.[1]
Furthermore, Britain, France and Germany are European nations with a strong spatial
planning tradition aimed at handling cultural landscapes.

The core dilemma of protected landscapes in Britain, France, and Germany, is that they are
no longer self-sustaining, and the links between landscape, community and economy no
longer self-reinforcing. Thus, the key issue for the future is what policy settings are needed
to ensure their survival in the face of environmental and cultural homogenization, as part of
the general process of globalization. In order to answer this question we discuss the
different governance strategies that are developed to re-couple socio-economic activity and
landscape quality in these protected landscapes. More and more, these strategies are co-
productions of public and private effort. This is a result of an ongoing shift in the above
mentioned state-society relations ('from government to governance') away from a top-down
approach towards more bottom-up approaches characterised by a decentralised style of
policy making that also stimulates the horizontal relations between public and private
bodies. Competencies are devolved to the regional level to allow for policy differentiation
and an administrative imperative to manage and control the public policy process to ensure
the achievement of national policy objectives in countryside areas.

The general aim of this chapter is to contribute to the recently started debate on sustainable
development of protected areas by comparing and assessing the different governance
strategies in British, French and German protected landscapes. This chapter starts with a
short introduction of the history and international context of landscape protection,
determining the particular western European experience with landscape preservation and
management. This brings us to the different landscape protection systems and strategies
adopted by Britain, France and Germany. We describe the identification and maintenance of
protected landscapes in these highly urbanized countries and analyze the forces that have
shaped them as well as the forces that are currently affecting the ecology and beauty of these
valued landscapes. Based on the comparison of the different protected landscapes, we
observe that attention for the potential of protected landscapes to stimulate sustainable
development is increasing. Despite the ubiquity of 'sustainability' as a concept, within
protected landscapes several attempts are made to protect the environment, to promote
sound development and to improve the quality of life for people now and in the future. In

---

[1] English geographer Aitchison (1995) has shown that the regions with the most intensive agriculture,
coinciding with the urbanised economic core region of Europe, are also the nations with the largest
perecentages of protected landscapes. It suggests that the protection of landscapes is less based on
biodiversity or on the degree of preservation of 'traditional agrarian landscapes', as it is on the values
and needs of an urban population.

the final section, some preliminary conclusions are drawn, and some remarks are given on the future of protected landscapes in Western Europe in a governance context.

2. Protected landscapes: History and international context

Protection of landscapes is not a recent invention. One of the historical landmarks was the designation of the first National Parks in the United States, Yosemite (1864) and Yellowstone (1872), aiming at the safeguarding of 'undisturbed' or 'primeval' nature (Runte, 1997). During the first decades of the twentieth century, a number of European countries followed, with Sweden, Switzerland, Spain and Italy in the front line (Hamin, 2002; Besio, 2003). Although the emphasis was on ecosystems that were seen as almost completely natural, gradually it became clear that all of them were in fact partly man-made landscapes. And even when very little human influence was recognizable, the national park designation itself defined these areas within the domain of human society (Mells, 1999). Therefore, the distinction between 'nature' and 'culture' became less strict. From the 1930s onwards a distinction developed between reserves that were protected mainly for their ecological values and a new group of old 'traditional' agrarian landscapes.

The densely populated character and the existence of little wilderness areas have, in contrast to North America, contributed to the fact that cultural landscapes have become an important management category in Europe (see Table 1). Conservation effort in most European countries has therefore focused upon agrarian, lived-in, working landscapes. These landscapes depend on human intervention. Since the European landscape is extraordinarily varied and rich in both natural and cultural interest, designation systems have been developed in order to protect the most beautiful and vulnerable parts. These protected landscapes, focused on the conservation of the specific uniqueness of cultural landscapes, lie at the heart of the identity of rural Europe and potentially enrich the cultural and natural diversity of both people and places (Pedroli et al., 2007).

	United States	Europe
Conservation of…	Wild, 'untamed' nature	Rural, lived-in landscapes
Status	Reserve	Protected landscape
Ownership	Public (state-owned)	Co-managed (Public-Private)
Type of area	Unoccupied	Inhabited

Table 1. Two types of Park Model

Against this background it is not strange to note that the European experience with protected landscapes is varied. Each country has taken a different course according to its geographic and historical characteristics, social structure, political organization and planning culture. As a result European protected landscapes show many differences, in the types and number of designated areas they have established, their legal structures, tasks, as well as in their proportion related to the countries surface. However, certain common characteristics can be identified. It almost always involves (rural) landscapes that are important for their traditional and less intensive land-use. In most cases these landscapes are inhabited by private land-owners (mostly farmers) with some small federal or state holdings and co-managed by public and private parties. Authority, responsibility and accountability for managing the protected landscape are shared in various ways among a

variety of actors like government agencies, local communities, non-governmental
organizations (particularly environmental groups) or private landowners.

Although the officially designated landscapes in Western Europe are often called national or
regional parks they are, according to international guidelines by IUCN (World Conservation
Union/International Union for Conservation of Nature and Natural Resources) defined as
Category V protected areas. IUCN (1994) defines protected landscapes (Category V) as "areas
of land, with coast or sea as appropriate, where the interaction of people and nature over time
has produced an area of distinct character with significant aesthetic, ecological and/or cultural
value, and often with high biological diversity". Category V areas represent only some 9% of
protected areas globally (6% by area). But in Europe, the UNEP-WCMC database records that
some 46% of the total area under protection is in Category V (Chape et al, 2003).

The disparity of landscapes that fall into Category V is substantial. The classifications
according to national law include, for instance, *Parco Naturale Regionale* (Italy), *Parc Naturel
Régionaux* (France), *Naturpark* (Austria and Germany), and *National Park* (Britain). Recently,
so-called *Geoparks* have been established in different European countries, with the specific
objective to protect geological heritage.[2] The perspectives of geological heritage conservation
of Geoparks are positioned within the frame of the wider and more complex strategy of
conservation of the natural and historical-cultural heritage that the territory presents, acting
through efficient management measures able to couple strategies of active protection with
actions aiming at the enhancement and the social-economical development, including
geotourism. Both Nature and Geoparks are a specific type of category V areas. They are
protected landscape areas, which have developed trough the interaction of man with nature.

Unlike the term 'nature' suggests, 'nature parks' are not managed for nature and
biodiversity purposes but for landscape conservation and recreation. Recreation and
amenity oriented purposes, but also culture and rural development, therefore, are mostly
dominant over the pursuit of nature conservation. Currently nature parks get worldwide
attention under the IUCN protected areas category V (see Table 2). They experience
attention due to their increasing attractiveness as areas of leisure and valuable habitats as
well as their less strict guidelines and planning objectives. Due to their central task to
connect protection and the use of cultural landscapes lastingly they are gaining significance
for the future. Only on the basis of continued use the cultural and geological heritage
landscapes in Europe and their large biodiversity can be secured in the long term (Schenk;
Hunziker & Kienast, 2007; Panizza, 2001; Farsani, 2011).

3. Western European approaches to landscape conservation

Throughout Western Europe more and more landscapes are maintained with the specific
aim of preserving the cultural landscape regarded as valuable by the (urban) society. These
protected landscapes (Category V) seem to be best supported by sustainable policy

---

[2] The Geoparks in Europe are part of a European Geoparks Network that was established in June 2000
and now consists of 37 Geoparks in 15 countries of the European Union. In February 2004 the European
Geoparks Network was formally integrated into the UNESCO-endorsed Global Geoparks Network. The
Global Geoparks Network, assisted by UNESCO, provides a platform of active co-operation between
experts and practitioners in geological heritage.

objectives and measures. The social conception generally considers these landscapes as patrimony; this seems appropriate because changes in traditional cultural landscapes have often been very slow, and they seem to be definitely stable and therefore an appropriate symbol of regional and national identity. We therefore argue that landscapes and the efforts to preserve them are never neutral or objective. The specificity of landscape and its meanings are first and foremost cultural. For instance, landscape is seen by national governments as an important national asset that contributes to national pride and identification (Lekan, 2004).

Initiative	Geographical scope	Type(s) of landscape	Policy perspective
World Heritage Convention (UNESCO)	Global	Landscapes of exceptional, universal importance	Conservation of natural and cultural heritage
Global Network of National Geoparks (UNESCO)	Global	Territories containing geology of outstanding value	Geological heritage and sustainable local development
European Landscape Convention (EU)	Europe	All landscapes: rural and urban, vernacular and extraordinary, designed and planned	Protection, management and development of landscape
Protected Areas (IUCN-Category V)	National/regional	Important agrarian/rural cultural landscapes	Sustainable development and reinforcement of natural and cultural values

Table 2. International perspectives on landscape. Source: Selman (2006); Farsani, Coelho and Costa (2011).

Since landscapes play an important role in building the national identity the origin of the preservation of landscapes is often rooted in processes of nation building. Landscape preservationists often promoted the cultural construction of nationhood by envisaging natural landmarks as touchstones of emotional identification, symbols of national longitivity, and signs of a new form of environmental stewardship. For instance, Olwig has shown that with the growth of the power of the state in the Renaissance, the concept of landscape as land and custom became subverted by the state. Landscape, as he argues, became the territory controlled by the state – embodied by the monarch – and made visible as scenery through theatrical and pictorial representations (Olwig, 1996; 2002). The view of landscape as scenery was later adopted by tourists and conservationists, and remains a dominating paradigm in current landscape management and administration by state and other public authorities throughout Western Europe.

Building on the ideas of Olwig, we argue that landscape conservation systems are shaped by socio-cultural patterns of perception and tradition. In order to understand the culturally and historically varied character of western European landscape protection it is necessary to reveal the connections between nation-building and landscape protection. In what follows, we highlight the evolution of different landscape conservation systems in modern Britain, France, and Germany against the background of the mutually reinforcing processes of nation-building, state intervention and planning peculiarities. The conservation systems we analyze are the British *National Parks*, French *Parcs Naturels Régionaux*, and German *Nature Parks*, all IUCN-Category V protected areas. Each conservation system is described: its objectives and results. Finally, each section concludes with a short overview of current governance strategies to deal with the co-ordination of various actors to pursue a more sustainable development of landscape.

3.1 British national parks

3.1.1 British conservation history

The idea that (national) identity, landscape and history are interlinked is nowhere as manifest as in Britain (Bishop, 1995). Responsible for the emergence of the national parks movement that led to the creation of the British National Parks, are the rapid urbanization, industrialization and agricultural rationalization during the first half of 19th century. In 1815 London had about 1, 5 million habitants, in 1860 3 million; figures and growth never seen before in world history. And London was not the only big city in the country. A stunning 25% of the whole population already lived in cities; urban sprawl was everywhere. The impact of this fast urbanization, and also of the main driving forces behind it, i.e. fast, agricultural rationalization and large scale, coal and steel based industrialization, was very visible everywhere. The effects on nature and the countryside were often very depressing, aesthetically, ethically, socially, culturally and ecologically, and so far reaching and fast that people felt alienated.

It was in this setting that the longing for 'natural' landscapes arose, in the form of nostalgia for a lost past, characterized by beauty, rurality, harmony, proportionality and cohesion. The pioneers of this new line of thinking and feeling were members of the urban elite; men like John Ruskin and William Morris. The obvious negative and certainly hideous effects of the fast agricultural rationalization, urbanization and industrialization shocked them. They called up to appreciate and respect the beauty of the land and criticized the prevailing purely utilitarian attitudes and practices. They stressed the value of social cohesion, and sought to bring it back it by restoring the relationship with the land, based on aesthetic criteria. According to Ruskin 'all lovely things are [...] necessary, the wild flower as well as the tended corn, the wild birds and creatures of the forest as the tended cattle; because man does not live by bread alone' (Ruskin, 1985, p. 226). Morris emphasized that the British people 'must turn [their] land from the grimy back-yard of a workplace into a garden' (Morris, 1969, p. 49-50). In doing so Ruskin and Morris expressed the feelings of a large and fast growing segment of the urban middle classes.

From the second half of the 19th century onward more and more citizens started to organize themselves in voluntary organizations, with the goal to preserve nature and culture. These organizations spread new ideas and ideals about the value of scenic beauty, rural live, cultural heritage and identity and their unbreakable bond with the British landscape, such as the idea of the countryside as the almost sacred locus of British identity, with its hamlets, forests, meadows, cottages and hedges. One of the main characteristics of this new attitude was a huge aversion to the degrading effects of industrialization and urbanization, and a tendency to give in to nostalgia and feelings of alienation and loss, emotions to be compensated by disappearing in the beauties of nature and the countryside. The emphasis was on beauty, on aesthetical, aspects, and the idea that the good and the beautiful went together and were to be found on the countryside and in nature, and the idea and the bad and the ugly were to be found in the city and industry.

The pre-war national parks movement drew its strength from the convergence of several traditions. There was the cause of protecting the most beautiful scenery that had its roots in the writing of Ruskin, Morris and Blake. But this strand of the national parks movement had a strong class bias and its leaders often feared, and sometimes opposed, the urban masses who wished to holiday in the Lake District for example. It thus contrasted with the democratic, even Marxist leanings of a second strand that was concerned with access, and the rights of the working man to enjoy the open moors and fells, principally around our northern industrial cities. The third strand behind the national parks movement was scientific; its origins can be traced back to the nineteenth century pioneers, like Charles Rothschild, the founder of the Society for the Promotion of Nature Reserves, and its aims were to ensure that nature conservation was placed on a statutory footing.

Only when these forces combined did they create a powerful political pressure for legislation, but it took the Second World War to create the conditions where such legislation could be enacted. Writing in 1947, Clough Williams-Ellis, the visionary who created Portmeirion, dedicated a book about the National Trust to all those beautiful natural and other places that had been destroyed during the war years – "a massacre of loveliness" he called it (William-Ellis, p. 7). Beauty was indeed the victim of wartime "collateral damage", inflicted daily on a huge scale around the country, and indeed across the world. The passions and outrage that this gave rise to among the public and the political elite, and the belief that the nation needed to offer its citizens a better physical environment after the war, made the famous 1949 National Parks and Access to the Countryside Act possible (see for a history: Sheail, 1975; MacEwen & MacEwan, 1987; Evans, 1992).

3.1.2 Centralized planning system

The National Parks and Access to the Countryside Act 1949 is an Act of the Parliament of the United Kingdom which created the National Parks Commission which later became the Countryside Commission and then the Countryside Agency, provided the framework for the creation of National Parks and Areas of Outstanding Natural Beauty (AONB) in England and Wales, and also addressed public rights of way and access to open land. Currently, 12 National Parks are designated, of which the South-Downs National Park is the last of the 12 areas, designated in March 2009. Their main goal is to conserve and enhance the natural beauty, wildlife and cultural heritage of the areas, in mostly poor-quality agricultural upland. Furthermore, since 2003 seven so-called Geoparks have been created in the UK. The first one was the North Pennines Geopark.

The British National Parks were set up in a system of heavy-handed centralized planning. Development control by the National Park Authorities (NPA), that is the detailed system by which approval is sought for building and land use change, is one of the main instruments of park management. Protective measures and financial resources are provided by central government. Because the adopted system manifested major policy performance problems in the 1970s and 1980s the traditional role of the NPAs in controlling development shifted to one of influencing land management (Curry, 1992). The management of land by the NPAs

has focused on mitigating the worst effects of the European Union Common Agricultural
Policy. This activity was largely reactive, seeking to swim against the tide of changes forced
on the Parks. Protection took place largely in isolation from, or frequently in opposition to,
the most important political pressures on rural life.

Because of the emphasis on development control British parks have alienated local farmers
and communities, whose cooperation is needed to carry out conservation policy. Therefore,
the 1991 Edwards' review of the British National Parks, *Fit for the future*, resulted in the
addition of the economic and social well-being duty in Section 62(1) of the Environment Act
1995. The Environment Act 1995 makes a move towards integrating functions in respect of
National Parks. The purpose of preserving natural beauty is extended to 'protect, maintain,
and enhance the scenic beauty, natural systems, and land forms, and the wildlife and
cultural heritage'. According to Edwards' review, Park Authorities should foster the social
and economic well being of the Park communities in partnership with those organizations
for whom this is the prime responsibility. Experiences in putting this duty into practice,
however, are mixed. A co-ordinate planning and partnership working in support of the
economic and social well being of park communities is lacking. The (financial) restrictions
imposed under Section 62(1) are not helping either.[3] Consequently, Park communities feel
that their interests are not served well enough.

3.1.3 Park planning and partnerships

In the particular and influential British tradition landscape planning has mainly been
concerned with an agenda of protection, preservation, amenity and ornament. This focus
has been important, but has remained peripheral to a wider agenda of sustainable
development. In the first part of the twenty-first century, however, landscape planning
seems to become identified more strongly with the core concerns of sustainable
development and spatial planning. Through innovations such as the European Landscape
Convention, landscape has become increasingly central to matters of sustainability and
place-making. Currently, National Parks are positioned as models for sustainable
development in the British countryside, and the National Parks are given money by the
national government to encourage individuals and communities to find sustainable ways of
living and working, whilst enhancing and conserving the local culture, wildlife and
landscape.

The British landscape preservation tradition and its cornerstones, the National Parks, is
opening up and hooked on debates about sustainable development across rural and urban
domains. However, the failure of socio-economic partnerships within the Parks is a major
stumbling block on the road to sustainable development. Since there is a need to seek a new
balance between the protection of the natural beauty and the stimulation of the socio-
economic needs of park communities, recent initiatives in Britain increasingly respond to
the challenge of sustainability in Category V protected areas. For instance, the newly
established Scottish National Parks (2002) are to promote sustainable social and economic

---

[3] Section 62(1) of the Environment Act states that NPAs "shall foster the economic and social well-being
of local communities within the National Park, but without incurring significant expenditure in doing
so, and shall for that purpose co-operate with local authorities and public bodies whose functions
include the promotion of economic or social development within the area of the National Park".

development of the area's communities, next to the conservation and enhancement of the natural and cultural heritage (McCarthy et al., 2002).

Furthermore, a recent review report of the Welsh National Parks calls for a more integrated sustainable development approach in order to ensure a sustainable future for the (Welsh) National Parks. The report recommends a new park purpose to "promote sustainable forms of economic and community development which support the conservation and enhancement of natural beauty, wildlife and cultural heritage of the areas" (Land Use Consultants, 2004: iv). In order to act upon these new proposals, British park planning and management must be carried out in close partnership with the local community, private sector and relevant government organizations. According to Phillips and Partington (2005) recent innovative policies in Wales already use protected areas as places where sustainable forms of rural development are pioneered and promoted, giving substance to the British National Parks' new purpose.

3.2 French parcs naturels régionaux

3.2.1 French conservation history

The origins of the French landscape conservation movement that led to the creation of the *Parcs Naturels Régionaux* can be traced back to the late 19th century, when French politicians and administrators in the capital city of Paris developed their ideas about a centralized nation-state (Alard et al., 1992). The overall aim of the famous French centralization efforts of the 19th and early 20th century was to remould all aspects of regionally bounded life, socially, culturally, politically and economically (Weber, 1976). The aspiration was to re-forge rural and village France with its small peasant farms, by destroying the benumbing diversity in regional languages and cultures, and create a new unity, a new 'imagined community', as Benedict Anderson has put it, by blending and sometimes inventing new identities, goals and preferences. A clear example is to be found in the explicit efforts to create the impression that there was and always had been an unique French identity, embedded in and symbolized by the French countryside and French farmer, a process very present in the work of the famous French historian Jules Michelet, 'the man who invented the idea of France' (Braudel, 1998), for instance in his *Histoire de France* (1883).

The explicit purpose of centralization and modernization was to destroy existing old local identities and cultures, in particular the strong and very old links between region, identity and culture. To mention just one example, all existing regions and 'pays' in France, some of which already existed since Roman times, were intentionally split up in new small administrative units: departments. The borders of those departments intentionally cut across pre-existing cultural and political borders. Before the modernization and centralization of rural and village France, there existed no such idea as a unified French identity; identity was locally bounded, so completely self-evident that there was no need to talk about it. Or to put it differently: rural populations had heretofore been in France but not of it. For most French peasants and farmers local identity was all encompassing, replicated in the daily activities, rooted in the natural environment, and mirrored by the cultural environment. It is no coincidence that the most common and oldest French word for farmer is 'paysan', and that for landscape is 'paysage'. Identity in France was that what connected farmer, landscape and country(side): paysan, paysage and pays. Until the late nineteenth

century many of the French people belonged, by language, outlook, and culture, only to
their rural pays; at most their frame of reference was the province.

The modernization of rural and village France became a relative success: the rural culture
was assimilated into the national culture, as well as regions and their *paysage*, and Paris
developed into a capital city, overruling all other cities. However, around 1950 the planned
socio-cultural and economic centralization had become so successful that their combined
outcome tended to turn into a problem. France had indeed become a one nation state, with
one broadly shared language, culture and identity. It also had become a nation completely
dominated by the city of Paris. In 1950 almost 5,5 million people, more than 10% of the
French population, lived in Paris, and the expectation was that this number would rapidly
increase in the near future. The capital and its direct vicinity thrived. Every economic,
political or cultural institution of any importance was located in Paris; every decision of any
weight was taken there. The dominance of the central city and the central culture was so
strong that the province, the other cities and other parts of the country, started to crumble,
demographically, economically and culturally. Therefore, the French government decided to
change course.

Post-war planning effort in France, known as the 'amenagement du territoire', attempted to
more evenly redistribute the French population across the country as a means, in part, of
boosting its economy. Particular growth regions were designated, new administrative units
bigger than the existing departments, evenly spread over the country. The intention was to
stimulate the economic growth of those regions, improve their accessibility and
attractiveness, and reduce the pressure on Paris. Motorways and high-speed rail would
connect these regions, with each other and Paris. Each region would have its own main
urban centre, with all the necessary services and cultural and natural facilities. This step was
the first one towards a more decentralized policy, the first time in decades that (some)
power was delegated back from central government to the regions.

Provincial and agricultural France, whose memory, cultural and landscape legacy was lost
in the centralization efforts of the 19th and early 20th century, was in many ways
rediscovered. The ambition to allow regions space to reclaim their own identity, and the first
hesitant steps to cautiously promote these regional identities, became visible in the idea to
establish so-called *Parcs Naturels Régionaux* (PNR), a concept formalized by law in 1967.
These parks were designated by the central government in selected regions, and had to
combine the protection of the valuable natural and rural patrimony with regional rural
development. The underlying inspiration was "to contribute, in line with the general policy,
to a better distribution of the population over the whole of the territory, and the human and
economical revitalization of the rural zones" (Minister André Fosset, June 11, 1976). So, in a
way, the parks were a plan-led effort to mitigate the negative side-effects of decades of
modernization and centralization, processes that themselves had been object of state-led
planning.

3.2.2 Bottom-up approach

From the beginning most regional parks employed very strict rules with regard to land and
property development and architectural styles. They became breeding grounds for
landscape architects and architects, specialized in 'critical regionalism' (Lefaivre & Tzonis,

2004). In 1987 the idea of sustainable development was introduced. This resulted in 1988 in a reformulation of the main objective of the parks, namely: "to protect and manage the natural and cultural patrimony, promote economic and social development, and function as examples and places for experimentation and research". However, it was only in 1993 that the establishment and mission of PNR was legally formalized. Their formal mission became: "to contribute to the policy of environmental protection, land use, economic development and social and public education ... for the preservation of landscapes and the natural and cultural heritage" (Article 2, Loi Paysages, 1993). Environmental, economic and social issues were seen as mutually dependent, as were the ideas of preservation and development, and those of cultural and natural heritage.

Lessons with community participation and co-production of public and private partnerships can be learned from the French *Parcs Naturels Régionaux* (PNR) with their dual purpose of (1) preservation of the natural and cultural patrimony; and (2) economic development through more efficient agriculture, recreation, local handicrafts, and tourism. The French areal protection system also distinguishes national parks; these however are focused on biodiversity and nature conservation. The French regional parks have a history of developing the countryside while at the same time protecting the environment. This is reflected in the PNR emphasis on 'conservation through appropriate development' as Dwyer (1991) has argued. However, in contrast to the British parks, the French PNR lacks strong regulatory and enforcement powers. Consequently, a 'bottom up' rather than a 'top down' system has been developed that actively engages local park communities and organizations in a cooperative manner. The French PNR do not provide specific legislation for environmental protection, but instead functions through local coordination of existing land-use regulations.

Each French PNR is governed officially by a Charter, a statutory instrument which sets out its goals, the strategy designed to achieve them and a broad outline of the supporting actions. A 'chartered authority', made up of representatives of local, regional and national government stakeholders, is responsible for implementing the Charter. Consequently, the Charter, a contractual document that is approved by several representatives of local and regional agencies and NGO's, signs up Park plans. Under the Charter, rural communities accept the obligation to apply constraints to them selves concerning the treatment of the environment (Lanneaux & Chapuis, 1993). The chartered authority, a so-called *Syndicat Mixte*, enjoys planning powers at the sub-regional level relatively similar to those held by the National Parks Authorities in Britain. It will draw up a ten-year action plan. When that period is up, a review procedure examines the parks past accomplishments and if the park merits renewal of its charter, the objectives for the next ten years will be agreed by the authority and endorsed by the relevant regional environment directorate.

3.2.3 Regional rural development

Although in the early years (1970s and 1980s) the French parks mainly emphasized economic development of disadvantaged rural regions, from the early 1990s onward a shift in attitudes away from rigid economic utilitarianism can be observed. Currently, the French PNR develop strategies that either seek directly to support local economic activities or stimulate new socio-economic benefits that strengthen local cultural and natural heritage. Therefore, PNRs adopt a multi-functional approach: protecting both biological and cultural

From Landscape Preservation to Landscape Governance: European Experiences with Sustainable
Development of Protected Landscapes

245

diversity, and with preserving special landscapes and geological heritage-sites, while implementing a programme of social and economic development. PNRs evolved from a rather introspective organisation dedicated almost solely to the protection of the natural heritage and traditional ways of regional life to an outward-looking body determined to utilise local assets and communities involvement to achieve its goals. Furthermore, park authorities give advice to towns and villages regarding urban organization and the insertion of buildings in the landscape. Underlying is the idea that environmental protection and economic development are not mutually exclusive. Even more so, it is believed that economic decline could be harmful to the protection of the valued landscape and heritage. After all, in the French context, rural depopulation and marginalization are serious threads. As Buller (2000) has argued, the PNRs have made 'local economic revitalization their central mission'.

Since the late 1990s the French central government has committed itself to the idea that PNRs are perfect units for sustainable policy making (FPNR, 2007). The PNRs play a key role in contemporary regional rural development by applying the principles of sustainable development. Although some regional parks fail to implement the conservation objectives of park Charters, comparative studies on the British and French system have shown that the French regional parks surpass the British national park system in achieving a balanced regional development (Dwyer, 1991). According to LaFreniere (1997) the Park Chartres have had a moderating effect on the scale enlargement and intensification of agricultural practices and, furthermore, contributed significantly towards raising the awareness of local park communities regarding environmental impacts of economic development. The Charter model used by all French PNR to set goals, draw up action plans and measure both outputs and outcomes has proved particularly useful to involve local communities and indigenous attributes and resources, rather than on attempting to import economic success from somewhere else.

In 2007 there were 45 regional nature parks in total, covering 12% of France, involving 21 regions and more than 3 million inhabitants, and about 5% of the population (Historique de Parcs Naturels Régionaux, 2007). The regional parks have become icons of French landscape planning, of the possibility to combine protection and conservation of nature, landscape, culture and local identity with rural economic development and tourism. The regional parks give regions identity and attractiveness. They are key eco-tourism attractions, for the French themselves and for foreigners. This great emphasis on historicity, locality and rurality, however, also has its drawbacks. It limits the scope of possible development and tends to stiffen planning efforts. The emphasis in French planning on the physical aspects of spatial identity intensifies this process. The emphasis on locality also easily prevents the emergence of supra-local planning, for instance the realisation of ecological corridors between parks, and it easily confines interest for sustainable or responsible landscape development to regional parks.

3.3 German nature parks

3.3.1 German conservation history

The German nature and landscape conservation movement, responsible for the German Nature Parks (*Naturparke*), was very much influenced by the concept of *Heimat*, home or homeland; a concept that – up until today – influences German society at large (Lekan, 2004). For instance, in 1984 the first eleven parts of a series called *Heimat*, written and directed by the German filmmaker Edgar Reitz, appeared on German television. The series was about the development of Germany (former Federal Republic) between 1919 and 1982. The successive members of a family, but above all their native region and village, their so-called *Heimat*, played the leading part. The series was about the tension between on the one hand the desire for identity, locality, security and belonging and, on the other hand, the craving for freedom, liberalization and cosmopolitanism.

The German concept of *Heimat* expresses a "feeling of belonging together" (Applegate, 1990). It has a connotation that roams somewhere between the French idea of *pays*, the English notion of *home* and the Dutch notion of *heimwee*. *Heimat* is about the myriad emotional ties that link up someone's identity with the identity of ones birthplace (i.e. home, village, and region), expressed by the landscape, nature, agricultural practices, handicrafts, dialects, people, history and customs of that place; in short: all the 'places, objects, practices and images' that generate and sustain those (nostalgic) emotions, in the first place the parental home and village. It refers simultaneously to a *état d'âme*, a sense of place, the place itself, and the objects and practices at that place.

In the late 19th century German people started to seek refuge in so-called *Agrarromantik* (dreams that glamorize rural live and the countryside). This trend was especially strong amongst the (new) urban middle classes, most notably amongst teachers, civil servants and the clergy (Bergmann, 1970). They developed a new vision on the good live, based on new ideas about belonging, wholeness, culture and identity; ideas that rooted in sentiments that opposed the city to the countryside, and the present to the past. They 'decided' that the heart of German identity was to be found in the *Heimat*, conceptualized out of a mixture of traditional pre-industrial rural regions, villages and landscapes. That (imagined) *Heimat* had to be taken care off, protected where that was needed, and restored where that was possible. Those ideas and sentiments were bundled by E. Rudorff in a new practice oriented concept, the Heimatschutz ('Protection of native country').

The motivation behind the Heimatschutz movement was based on emotions, ethics, and aesthetics (Rollins, 1997). The aim of Heimatschutz was to explicitly protect, study and strengthen the Heimat, in all its aspects. One important component, in fact a cornerstone, was the protection of the countryside and it's history-rooted customs, practices, architecture and landscapes: the parental country, 'home of the German soul'. This ambition was not to be taken lightly; it went beyond pure aesthetical considerations, as a Saxon Minister articulated strikingly in 1915: 'Heimatschutz is no game, but rather a far-reaching cultural movement, whose influence pervades every corner of the nation... no more and no less than the preservation and re-creation of the basis of all culture: the raising of the feeling of Heimat, the protection of beauty and of historical uniqueness, the artistic education of people to good taste, and thereby also the raising of the economic power of our people'.

In 1904, a number of associations dedicated to these conservation ideals merged to form the "Bund Heimatschutz" (homeland conservation alliance). It was difficult to achieve contextual unity and solidity within the alliance, one reason being the often regional and landscape-related self-conception of the member associations, and the alliance therefore became an umbrella organization. The nature conservationist groups split off in the mid-

From Landscape Preservation to Landscape Governance: European Experiences with Sustainable
Development of Protected Landscapes

247

1920s as they felt that their ideals were always seen as a mere partial aspect at the conferences of the Heimat and historic monument conservationists. Heimatschutz and nature conservation moved even further apart after the First World War, yet the concerns of both movements were accounted for in the Weimar Constitution of 1919 and both historic monument preservation and nature and landscape conservation were adopted as national objectives. Over time, both the representatives of Heimatschutz and of nature and landscape conservation became receptive to the antidemocratic, racial and nationalist movements in the 1920s and 1930s, allowing themselves to be monopolized through legal measures and more or less adhered to the ideologies of National Socialism.

Heimatschutz remained separated from nature and landscape conservation when work recommenced after the Second World War. Nature protection itself also witnessed a drawback because of the war and the following period of reconstruction. In 1950s and 1960s, however, both conservation bureaucracy and private groups, in particular the Nature Park Society (Verein Naturschutzpark, VNP, established in 1909 in Munich) led by Hamburg millionaire Alfred Toepfer, and the German Council for Land Cultivation (Deutscher Rat für Landespflege, DRL, established in 1962), presided over by Swedish-born Count Lennart Bernadotte, promoted the extensive conservation of nature and landscapes in German-speaking regions. On 6 June 1956 in the former capital city of Bonn at the annual meeting of the Nature Reserve Association, the environmentalist and entrepreneur, Toepfer, presented a programme developed jointly with the Central Office for Nature Conservation and Landscape Management and other institutions to set up (initially) 25 Nature Parks in West Germany. Five percent of the area of the old Federal Republic of Germany was to be spared from major environmental damage as a result. In the following years, the *Verein Naturschutz Park* won state and federal (financial) government support and different regional and local governments set up nature parks (Ditt, 1996).

For Toepfer, patron of Germany's nature parks, life and love of the outdoors was part and parcel of his combat against the perceived ills of modern society (Toepfer, 1957). Obviously, the pre-war Heimatschutz movement influenced Toepfer's view on nature and landscape conservation. The expanding cities of West Germany and their population had to be given space for recreation and leisure activities (walking, cycling, water sports, etc.). Furthermore, nature parks, had to provide opportunities for people to come face to face with nature. The ideal of Toepfer was to establish recreational 'oases of calm' in idyllic rural settings to offset the 'mechanization' of daily life in 'denatured' cities (Chaney, 2008). But as federal and state governments devoted more resources to spatial planning at the end of the 1950s, the nature park program also became a planning project overseen by technocratic experts who could settle competing claims on German space by multiple parties.

In the 1960s and 70s regional planners' involvement with nature parks forced socials conservationists like Toepfer to view nature parks not merely as scenic landscapes for rejuvenation but as "model landscapes" that might illustrate how to use the country's territory more efficiently and equitably. The emphasis on "model landscapes" was strengthened in the 1970s with the emergence of the ecology movement and the Green Party. As a result, the German nature parks, once commenced from a predominantly conservative, often nationalistic (Heimatschutz) cause, gradually became associated more clearly with the political left and the international movement to protect the global environment, though without losing its traditional base of support among social conservatives (up until today the sponsors of nature parks are usually clubs or local special purpose associations) and without completely abandoning its critique of modern civilization.

3.3.2 Protection trough usage

As shown, the original and central idea of Nature Parks was man's encounter with nature, the experience of the beauty of nature and scenery and the equal value of nature conservation and recreation. As was the case with the British national parks, German nature parks were mainly associated with public recreation. Emphasis solely was on stimulating public access of the German countryside, for instance by setting up visitor information centers. In keeping with this central idea, the tasks of landscape-based recreation were initially in the foreground: reasonable control of the increasing number of visitors, recreational facilities compatible with nature, and resolution of the conflict between nature conservation and recreation. The socio-political aspect of nature parks – to provide opportunities for recreation, especially for city-dwellers – was considered very important too.

Although the parks were popular and had a positive image, nature conservationists and environmental groups lamented that they were poorly administered, since few restrictions were placed on use (farming and forestry were permitted). Furthermore, nature areas were inadequately protected. As a result, conservation goals got more important, especially since the introduction of the 1976 *Bundesnaturschutzgesetz* (Nature conservation law) which gave the nature parks a legal status. The definition of the category of Nature Park was laid down in federal law (§ 27 of the BNatSchG). Paragraph 27 of the BNatSchG determined that natural parks are large areas that are to be developed and managed as a single unit, that consist mainly of protected landscapes or nature reserves, that have a large variety of species and habitats and that have a landscape that exhibits a variety of uses. Basically all actions, interventions and projects that would be contrary to the purpose of conservation are prohibited. Nature parks are to be considered in zoning and must be represented and considered in local development plans. This is called an acquisition memorandum. They are binding and cannot be waived because of a higher common good.

From the late 1970s onward the aim of Nature Parks is to strive for environmentally sustainable land use. The underlying idea is "protection through usage". Self-evidently, the acceptance and participation of the population in the protection of the cultural landscape and nature is very important. In doing so the nature conservation and the needs of recreation users are linked so that both sides benefit: sustainable tourism with respect for the value of nature and landscape is paramount in today's Nature Parks. It was also in the late 1970s that management authorities were installed, trying to stand up for the best interests of the areas. Since then the regulation of the German Nature Parks are organised as a special purpose association (*Zweckverband*). However, they have been dominated by, for example, agricultural associations who opposed against land use regulations that would endanger their idea of agricultural modernization. Since 1995, following updated legislation and responses to international calls for sustainable development, most notably the Rio summit in 1992, as well as the reunification of West and East Germany, there has been a change in orientation towards much more active involvement of local stakeholders in the management of Nature Parks (Stoll-Kleemann, 2001).

3.3.3 Model landscapes

About 97 Nature Parks now cover about 25 % of Germany's area. They play a forward-looking and important role in the protection of nature, landscape-based recreation and the conservation of Germany's cultural and geological important landscapes. Their contribution is therefore decisive for the identity, preservation and development of the regions. Since the late 1990s there is a growing governmental interest in the conservation and recreational use of Germany's Nature Parks. This attention has to do with a shift to a post-productivist rural policy, as well as with a renaissance of cultural and natural heritage issues, like regional identity. As a result, most Nature Parks are subject of special funding from the federal government. This money is used to cover the purchase of agrarian land, to fund special conservation measures, and as compensation for limitations of existing land use. In addition, money from the state government (Länder) is geared to funding particular conservation contracts with farmers to maintain cultural and natural heritage.

The German federal state currently sees Nature Parks as "model landscapes" with their aim to preserve unique landscapes for and with man and to contribute to a sustainable regional development (Deutscher Bundestag, 2007). Therefore, the Association of German Nature Parks (*Verband Deutsche Naturparke* [VDN]) is supporting Nature Parks in correspondence to their tasks by law in the promotion of an environmentally friendly and sustainable tourism, in the establishment of an ecological land use, which protects and recovers biodiversity and in proceeding regional development, which is maintaining cultural landscapes (VDN, 1995). To widen the possibilities of environmental education for visitors and the local population therefore is another task the Association, together with the help of the different park authorities, takes care of.

In the parks emphasis is being placed on promoting regional agricultural and forestry products and tourism services and in this way encouraging appropriate variants of land use. In addition to nature and landscape conservation, German natural parks also play an important role in preserving local customs, traditional crafts, historical settlement patterns, and regional architecture. Different projects, therefore, attempt to guarantee the economic advantages deriving from rural economic renewal and the advantages of a rediscovered sense of regional identity. The management philosophy of most Nature Parks embraces the peaceful coexistence of nature conservation with sympathetic economic enterprise and sustainable use of natural resources.

4. Landscape conservation and sustainable development

4.1 Converging conservation strategies

As the previous paragraphs shows, the origins, objectives and management of landscape protection systems throughout Western Europe differ significantly (see Table 3). In the Britain the case was, first and foremost, to conserve the most spectacular, wild or geomorphologic valuable landscapes by establishing National Parks. The establishment of National Parks reflected a particular aesthetic tradition, that was influenced by writerly and artistic conventions, and was applied to areas agreed by a relatively like-minded community of campaigners. It also affirmed the notion of British landscape as something which could be framed and separated from its less worthy surroundings. In France the main goal was to enhance rural development in fragile but interesting cultural or geological landscapes. The establishment of PNRs was influenced by the particular French tradition of territorial planning and affirmed the notion of the landscape as something that could strengthen regional identity. In Germany, at last, Nature Parks were conceptualized as antidote for an urbanizing society longing for leisure space. The establishment of *Naturparke* was inspired by social conservationist thinking, and idealized a rural Germany, which had to be rediscovered ('Heimat neu entdecken').

	Britain	France	Germany
Category	National Parks	Parcs Naturel Régionaux	Naturpark
Objective	Protection of landscape, stimulating outdoor recreation	Conservation of cultural or geological heritage and stimulating of rural economy	Sustainable development of the countryside to protect and enhance nature and valuable landscapes
Number (in 2009)	14	45	97
Area (of total country)	10%	13%	25%
Administrative organisation	Park Authority	Syndicat Mixte	*Zweckverband* (Special Purpose Association)
Preservation	Development control	Sectoral legislation and cultural history	Landschaftsschutzgebiete (landscape protection areas)
Management	Conservation contracts Zoning of Land Use Protected areas Ecosystems services Branding	Landscape contracts Ecomuseums Education Regional products Architectural restrictions Branding	*Wettbewerbe* (contests) Conservation contracts Eco-Tourism Regional products and crafts markets Branding
Finance	National government Heritage Lottery National Trust European Funds	Municipalities Regional governments Civil society organisations European Funds	Municipalities Kreise (regional governments) European Funds

Table 3. Protected landscapes (nature parks) in Britain, France and Germany. Source: Janssen et al. (2007)

Although different in (cultural and historical) origin and objective, recent policy proposals for protected landscapes in Britain, France and Germany converge towards a broadened sustainable development perspective. The original (pre-ecological) idea of protected landscapes as synonymous with scenery, farming as a protector rather than industrialiser of the countryside, and a system of enhanced (spatial) planning controls to safeguard the environment became outdated and obsolete. The narrowly preservationist concept, focused on applying measures necessary to sustain the existing form, integrity, and materials of landscape, gradually evolves into a more inclusive and social view of conservation that links nature and culture, protection and development, top-down planning and bottom-up approaches. It is recognized that designated landscapes are essentially evolving, changing, with new layers continually being superimposed on older ones. This is true for natural change, even more so for change caused by human impact. Human beings have shaped and changed the landscape they live in. The cultural landscape cannot stay the same, as culture means action, experience, experiment, progress, and change.

Since the late 1990s British National Parks, French Parcs Naturel Régionaux and German Nature Parks have begun to serve a wider set of social, economic, geological and ecological purposes, including, for instance, adressing quality of life, climate change, conservation of biodiversity, and protecting cultural and geomorphosite heritage. The apparently unbreakable relationship between landscape and visual matters, such as 'scenery' and 'aesthetics', is, therefore, forced open. Obviously, landscape in these modern parks means more than just a scene appealing to the eye. Increasingly, landscape is used as a holistic concept around which a wide array of disciplines can coalesce to explore the integration of human-nature relationships. Furthermore, a shift has taken place in the governance approach of these protected landscapes. Governing of protected landscapes in Britain, France, and Germany more and more relies on networks of interconnected actors from the public, private and voluntary sectors rather than a hierarchy dominated and defined by the central government.

Today's governance of protected landscape designations in Britain, France, and Germany takes place in partnership with those who work and live in the landscapes. Local communities are engaged in the enjoyment, understanding and stewardship of the cultural landscape. Partnerships are set up by the governing park authorities (National Park Authorities, Syndicat Mixte, Zweckverband) in order to build capacity, especially in the commercial and voluntary sectors, to ensure that in the long-term there is the critical mass of skills and expertise needed to sustain informed conservation of the natural and cultural heritage in protected landscapes. The devolving impulse of the British central government, for instance, has resulted in a growing awareness on the part of the Park Authorities of the potential benefits of action in (regional) partnership with local actors. NGO actors, businesses and private parties are involved in setting up landscape management strategies. Partnerships are seen as the key to successful implementation, with the different Park Authorities acting primarily as an enabler for sustainable (regional) development, undertaking or commissioning work where its skills and expertise, or its national and/or regional remit, will make the critical difference.

4.2. Living models of sustainable use

Throughout western Europe, and most notably in Britain, France and Germany, development of protected landscapes is no longer seen as a threat, to be repulsed by an additional layer of planning bureaucracy or authority. The acceptance of the paradigm of development, of course, is stimulated by a number of trends, such as urbanization and the rapid growth of outdoor leisure, the post-productivism of the rural sector, shifting state-society relations, as well as new insights in conservation science (ecology, geoparks) and spatial planning (multiple-use theories). Infusing all these trends is the emergence of the sustainable development discourse, popularised in 1980 by the World Conservation Strategy (IUCN et al., 1980), and firmly established in 1992 by the United Nations Conference on Environment and Development. By the close of the twentieth century, all areas of nature and landscape policy are being expected to demonstrate their contribution to more sustainable living.

The re-emergence of landscapes as cultural action arenas for sustainable development is inseparable linked to the dual process of globalisation and regionalisation. In the final quarter of the twentieth century the solidification of the concept of the nation-state and its unwieldy structure has been weakened, and with it the (homogenizing) notions of modernity and universality (Harvey, 1989). Local and regional specificities of space, form and place (territorial distinctiveness) are put forward to counteract the dislocation and lack of meaning in modern society. Contextual forces are to give a sense of place and meaning in a globalizing world. Increasingly cultural landscapes are seen as such a contextual force. After a period of nationalism we observe a renewed interest in the region all over Europe: regional differences and traditions are cherished, the issue of regional identity is widely debated, and new regional movements are emerging (Keating, 1998). Some even speak about the 'rise of regional Europe' (Harvie, 1994). The spatial and material dimension of this 'regional Europe' is symbolised by the manifold European cultural landscapes. The outstanding richness, regional diversity and uniqueness of landscapes form collectively a common European natural and cultural heritage (Pedroli et al., 2007). The existence of specific regional identities, each with its typical landscape heritage, is actively promoted, defended and helped by EU policy, programmes and funds, like the LEADER Rural Development programmes, INTERREG, and networks like the European Geoparks Network, built up with the support of European Union initiatives.

As a result of the emerging sustainability agenda a commitment to maintain and enhance the landscape quality of rural and urban areas is a central theme of several state and European visions of a sustainable countryside. Against this background protected landscapes throughout Europe more and more function as flagships for a new and integrated public policy for rural areas. Since landscape conservation and countryside development are aspects of a single whole, conservation increasingly is seen as an integral part of sustainable management. This is highlighted by the above-mentioned British, French and German protected landscapes (be it National Parks, Nature Reserves, protected landscapes or Geoparks), which, since the 1990s, strive towards a regional integration of agriculture, nature and landscape, thereby overcoming the often-strong sectoral division of countryside, regional and landscape policy.

Already in the 1980s IUCN recognized protected landscapes as "living models of sustainable use" (Lucas, 1992). Recent political commitment to sustainable development on a European level further strengthens the idea of an inclusive approach for protected landscapes (Council for the EU, 2006). The concept of sustainable development encourages policy officials to address the environmental and social as well as economic dimensions of rural areas. Because of the particular origin and nature of protected landscapes, principally the close relationship between landscape and the people connected with it category V protected areas [...] could very well "become pioneers in society's search for more sustainable futures" (Phillips, 2002). Several public policies in Europe have recently recognized the role of landscape within the framework of sustainable development. The following objectives have accordingly been articulated: regional policy – balanced opportunities for economic development and the provision of services; agricultural policy – compliance with environmental standards, cultural landscape preservation and multi-functionality; transportation policy – assignment of a high priority to railways and public transport; spatial development – rational use of space and the preservation of natural resources; environment and nature conservation – improved quality of the human environment, and the conservation of biodiversity and geomorphologic diversity.

As demonstrated by the European Landscape Convention (ELC) landscapes are more and more recognized "as essential components of people's surroundings, an expression of the diversity of their shared cultural and natural heritage" (Council of Europe, 2000: 4). The ELC argues that landscape should be valued for reasons of health, education and rural development. The Convention aims to promote landscape protection, management and planning, and to organize European co-operation on landscape issues. In the light of the perceived acceleration of landscape change, it seeks to "respond to the public's wish to enjoy high quality landscapes and to play an active part in the development of landscapes". Signatories to the Convention undertake to establish and implement landscape policies aimed at protecting, management and planning; to integrate landscape in the wider context of sustainability. By taking into account landscape, culture and nature, biotic and abiotic, the Council of Europe seeks to protect the quality of life and well-being of Europeans from a sustainable development perspective (Council of Europe, 2006).

Despite the ubiquity of 'sustainability' as a concept, within protected landscapes several attempts are made to protect the environment, to promote sound development and to improve the quality of life for people now and in the future. The principles of sustainability, for instance, are applied in a diversity of grassroots projects in order to stabilize and reduce the region's footprint. The intention is not to strive for a zero-growth situation but instead adopt a strategy that develops a mutual compatibility between environmental protection and continuing environmental growth. An interesting question in that regard is to what extent the emerging 'sustainability paradigm', which integrates economic activity with conservation in a sustainable manner, is running the risk of going too far in compromising conservation in favour of developmental interests (Antrop, 2006; Mose, 2007; Janssen, 2009).

5. Concluding remarks

This chapter has highlighted that cultural landscapes are increasingly understood as something not merely to be protected and preserved. The World Heritage Convention and the European Landscape Convention (Council of Europe, 2000, 2006) as well as the new concepts and strategies for nature parks in Britain, France and Germany propose considering cultural landscapes in general, and protected landscapes in particular, also as a force to promote sustainable (regional) development. The notion of development and change is a key component of the concept of sustainable development itself. Indeed, sustainable development not only involves sustaining what has been realised as Brundtland defines, but also sustaining future development (Brundtland, 1987). It means the preservation of opportunities, but also the creation of new resources and opportunities for future generations.

In order to realize sustainable territorial development, the emphasis in protected landscapes is shifting from maintenance to development. As a result, landscape conservation strategies not only protect cultural and natural heritage of cultural landscapes, but also enhance territorial dynamics that strengthen and requalify the (weakened) territorial assets, such as (regional) identity and nature. Sustainability – and thus the challenge for protected landscapes - is increasingly positioned in the character of change itself and not in terms of any optimal state, pattern or blueprint. Common historical roots, special landscape features, typical products, cultural traditions, as well as innovative projects are possible initial points for identity-based processes. In connection with governance arrangements cultural landscapes can be constituted as action arenas for sustainable development. As a result, cultural landscapes are not only public interest goods and services that directly affect the social well-being of individuals but also represent important urban and rural development assets. Cultural landscapes are part of a region's capital stock and base for the development of countryside communities.

Given the limitations of our current institutions to respond to landscape-scale change, landscape governane will require a high degree of collaboration to bridge disparate sectors, to integrate complex institutional layers, and to engage a wide array of actors in the sustainable development of cultural landscapes (Görg, 2007). Since multi-sectoral and multi-level partnerships are essential to an inclusive and participatory approach to landscape conservation, the intention is to stimulate and integrate mutual gains between sectoral interests by a 'conservation through development' approach. By working cooperatively with local and regional stakeholders, local, regional and national governments try to increase regional wealth creation, giving greater importance to rural areas, and creating more acceptance for landscape conservation among the local population and increasing awareness of nature and the environment among visitors.

Building multi-sector and multi-level partnerships for sustainable development of protected landscapes, however, is not an easy task for protected landscape authorities and institutions. Considerable conflict and opposition can easily arise. Most often causes of resistance have less to do with possible economic losses to local livelihoods arising from designation, but rather lie in the manner of consulting and involving local interests. Participation processes are often too late, too formal, and too narrow in compass. In addition, there can also be much miscommunication and misunderstanding between landowners, farmers, businesses and residents on the one hand, and the landscape conservation officials and experts on the other. Governance experiences with protected landscapes in Western Europe, therefore, emphasize the importance of communication skills, and capacity to create consensus among those who live and work in protected landscapes, to reduce scepticism and suspicion regarding the purpose of landscape conservation (Thompson, 2003, 2006; Janssen et al., 2007). It is only via the process of collaboratively acting together that full understanding and co-operation is achieved (Healey, 2007). Involvement and building capacity is key to securing sustainable stewardship of cultural landscapes (Selman, 2001).

We assume that governance for sustainable development of protected landscapes remains a challenging task in the 21st century. In that regard it is gratifying to note that there is an emerging (academic) debate on the influence of protected landscapes on local and regional development (Mose, 2007). Both in the academic debate and in conservation practice protected landscapes are recognised as keystones for sustainable development initiatives. National parks, Geoparks, eco-museums and landscape parks are unique constellations of 'nature', people, heritage, tourism and culture. These resources are managed with under appreciated pools of drive and expertise. Such areas demonstrate the real meaning of sustainable development, whilst conserving the exceptional natural and cultural heritage. We have attempted to contribute to the emerging (albeit under-theorised) area of protected landscapes within academic discourse by comparing British, French, and German landscape conservation approaches. However, given the large number of protected landscapes in Western Europe, and their increasing responsibilities in wider city and countryside development programmes, we think there is scope for more large-scale and in-depth (comparative) studies. Fortunately, a diverse range of initiatives is currently developed, focusing on a European-wide landscape research and action programme, substantially funded with a strongly integrative perspective. For instance, under the umbrella of UNISCAPE (European Network of Universities for the Implementation of the European Landscape Convention) professional networks are created to exchange information and expertise on landscape conservation and development (see: http://www.uniscape.eu/). These networks are essential to encourage and establish new and widely-shared approaches (including theories, concepts and methods) that will support more integrated, sustainable and socially-relevant landscape research as well as landscape management practices.

6. References

Allard, D. & Frilieux, P.N. (1992). Designation of and institutional arrangements for protected landscapes in France (a review and case study in Normandy). Council of Europe Conf. Protected Landscapes: A comparative European Perspective. UK.

Antrop, M. (2006). Sustainable landscapes: contradiction, fiction or utopia? *Landscape and Urban Planning* Vol. 75, 3-4, pp. 187-197

Applegate, C. (1990). *A Nation of Provincials: The German Idea of Heimat.* Berkeley: University of California Press.

Besio M., editor (2003). Conservation planning: the European case of rural landscapes. In: *Cultural landscapes: the challenges of conservation.* Paris (France): UNESCO World Heritage Centre, 60–68.

Bergmann, K. (1970). *Agrarromantik und Grosstadtfeindschaft.* Köln: A. Hein.

Beunen, R.; Opdam, P.F.M. (2011). When landscape planning becomes landscape governance, what happens to the science? *Landscape and Urban Planning* 100 (4). - p. 324 – 326

Bloemers, T.; Kars, H.; Van der Valk, A. (2010). *The Cultural Landscape & Heritage Paradox. Protection and Development of the Dutch Archaeological-historical Landscape and its European Dimension.* Amsterdam: Amsterdam University Press.

Borrini-Feyerabend G. (2004). *Governance of Protected Areas – Innovation in the Air.* (Retrieved from: http://www.earthlore.ca/clients/WPC/English/grfx/sessions/PDFs/session_1/ Borrini_Feyerabend.pdf)

Brundtland G. (1987). *Our Common Future, Report of the World Commission on Environment and Development.* Rome: World Commission on Environment and Development [Published as Annex to General Assembly document A/42/427].

Buller H. (2003). *The French Parcs Naturels Regionaux: Socio-Economic impact and rural development*. Newcastle upon Tyne: Centre for Rural Economy (working paper 52)

Chaney, S. (2008). *Nature of the miracle years: conservation in West-Germany, 1945-1975*. New York: Berghahn Books.

Chape S, Blyth S, Fish L, Fox S, Spalding M. (2003). *United Nations list of protected areas*. Gland: IUCN.

Council of Europe (2000). *European landscape convention*. Florence (Italy): Council of Europe.

Council of Europe (2006). *Landscape and sustainable development: challenges of the European Landscape Convention*. Strasbourg (France): Council of Europe.

Curry N. (1992). Controlling development in the national parks of England and Wales. *Town Planning Review* 2, 107–121.

Dearden, P.; Bennett, M. & Johnston, J. (2005). Trends in Global Protected Area Governance 1992 –2002. *Environmental Management* Vol. 36, No. 1, 89-100

Ditt K. (1996). Nature conservation in England and Germany 1900-70: forerunner of environmental protection?. *Contemporary European History* 5, 1–28.

Dowling, R.K. (2011). Geotourism's Global Growth. *Geoheritage*, 3, 1-13.

Dwyer J. (1991). Structural and evolutionary effects upon conservation policy performance: comparing a U.K. National and a French Regional Park. *Journal of Rural Studies* 3 (7), 265–275.

Farsani, N., Coelho, C. and Costa, C. (2011). Geotourism and Geoparks as Novel Strategies for Socio-economic Development in Rural Areas. *International Journal of Tourism Research*, 13, 68–81.

Freniere GF. (1997). Greenline parks in France: Les Parcs Naturels Regionaux. *Agricultural Human Values* 4 (14):337–352.

Fuertes-Gutiérrez, I. & Fernández-Martínez, E. (2010). Geosites Inventory in the Leon Province (Northwestern Spain): A Tool to Introduce Geoheritage into Regional Environmental Management. *Geoheritage*, 2, 57-75.

Fürst, D.; Lahner, M.; Pollermann, K. (2006). Entstehung und Funktionsweise von Regional Governance bei dem Gemeinschaftsgut Natur und Landschaft - Analysen von Place-making- und Governance-Prozessen in Biosphärenreservaten in Deutschland und Großbritannien. In: *Beiträge zur räumlichen Planung*, H. 82, Hannover.

Görg, C. (2007). Landscape Governance – The "politics of scale" and the "natural" conditions of places. *Geoforum* 38, 954-966.

Hamin E. (2002). Western European approaches to landscape protection: a review of the literature. *Journal of Planning Literature* 16: 339–358.

Harvey, D. (1989). *The condition of postmodernity: an enquiry into the origins of cultural change*. London: Blackwell Publishers.

Harvie, C. (1994). *The rise of regional Europe*. London & New-York: Routledge.

Healey, P. (1997). *Collaborative Planning. Shaping Places in Fragmented Societies*. London: Macmillan Press.

Janssen J, Pieterse N, Van den Broek L. (2007). *Nationale landschappen. Beleidsdilemma's in de praktijk*. Den Haag/ Rotterdam (The Netherlands): RPB/NAI Uitgevers.

Janssen, J. (2009). Sustainable development and protected landscapes: the case of The Netherlands. *Journal of Sustainable Development & World Ecology* 1 (2009) 37-47.

Jones, C. (2008). History of Geoparks. *Geological Society*, London, Special Publications 2008, v. 300, 273-277.

Keating, M editor (1998). *The New Regionalism in Western Europe. Territorial Restructuring and Political Change*, London: Edward Elgar

Land Use Consultants and Arwell Jones Associates (2004). *Review of the National Park Authorities in Wales*. Bristol: Land Use Consultants.

Lanneaux M, Chapuis R. (1993). Les Parcs Regionaux francais. *Annals of Canadian Geography* 573, 519–553.

Lefaivre, L & Tzonis, A (2004). *Critical Regionalism. Architecture and Identity in a Globalizing World*. Munich: Prestel Verlag

Lekan, T. (2004). *Imagining the Nation in Nature: Landscape Preservation and German Identity, 1885-1945*. Cambridge, Mass.: Harvard University Press.

Lucas PHC. (1992). *Protected landscapes – a guide for policymakers and planners*. London: Chapman and Hall.

MacEwan A, MacEwan M. 1987. *Greenprints for the countryside?. The story of Britain's national parks*. London: Allen & Unwin.

McCarthy J, Lloyd G, Illsley B. (2002). National Parks in Scotland: balancing environment and economy. *European Planning Studies* 5: 665–670.

Mels, T. (1999), *Wild Landscapes: The Cultural Nature of Swedish National Parks*. Lund: Lund University Press.

Morris, W. (1969). *The unpublished lectures of William Morris*. Detroit: Wayne State University Press.

Mose, I (Ed.) (2007). *Protected areas and regional development in Europe: towards a new model for the 21st century*. London: Ashgate.

Olwig, K.R. (1996). Recovering the substantive nature of landscape. *Annals of the Association of American Geographers*, 86, 630–653.

Olwig, K.R. (2002). *Landscape, Nature and the Body Politic – From Britain's Renaissance to America's New World*. Madison: University of Wisconsin Press.

Panizza, M. (2001). Geomorphosites: Concepts, methods and examples of geomorphological survey. *Chinese Science Bulletin*, Vol. 46, Supplement 1, 4-5.

Pedroli B, Doorn A van, Blust G de, Paracchini ML, Wascher D, Bunch F (2007). *Europe's living landscapes. Essays exploring our identity in the countryside*. Zeist: KNVV.

Phillips A. (2002). *Management guidelines for IUCN Category V Protected areas: protected landscapes/seascapes*. Gland/ Cambridge: IUCN.

Phillips A, Partington R. (2005). Protected landscapes in the United Kingdom. In: Brown J, Mitchell N, Beresford M, editors. *The protected landscape approach: linking nature, culture and community*. Gland: The World Conservation Union. p. 119–130

Powell, J, Wragg, A, and Selman, P. (2002). Protected areas: reinforcing the virtuous circle, *Planning Practice and Research*, 17(3), 279-295

Reynard, E. (2005). Géomorphosites et paysages. *Géomorphologie : relief, processus, environnement*, 2005, 3, 181-188

Reynard. E. & Panizza, M. (2005). Geomorphosites: definition, assessment and mapping: an introduction. *Géomorphologie: relief, processus, environnement*, 3, 177-180.

Rollins, W.H. (1997). *A greener vision of Home: cultural politics and environmental reform in the German Heimatschutz Movement*. Michigan: University of Michigan Press.

Runte, A. (1997). *National parks: the American experience*. Lincoln: University of Nebraska Press.

Ruskin, J. (1985). *Unto this last and other writings*. London: Penguin Publishers.

Selman, P. (2001) Social capital, sustainability and environmental planning. *Planning Theory and Practice*, 2(1), 13-30.

Selman, P. (2006). *Planning at the landscape scale*. London: Routledge.

Schenk, A., Hunziker, M. & Kienast, F., (2007). Factors influencing the acceptance of nature conservation measures—A qualitative study. *Switzerland Journal of Environmental Management* 83, 66-79.

Sheail, J. (1975). The concept of National Parks in Great Britain. *Transactions of the Institute of British Geographers* 66, pp. 41-56.

Stoll-Kleemann, S. (2001). Barriers to nature conservation in Germany: A model explaining opposition to protected areas. *Journal of environmental psychology* 21.4 (2001), pp. 369-385.

Stoll-Kleemann, S. & Welp, M. (Eds.) (2006). *Stakeholder Dialogues in Natural Resources Management: Theory and Practice*. Berlin-Heidelberg: Springer Verlag/Environmental Science.

Thompson, N. (2003). *Governing National Parks in a Devolving UK*. Leeds: University of Leeds.

Thompson, N. (2005). Inter-institutional relations in the governance of England's national parks: A governmentality perspective. *Journal of Rural Studies* 21, pp. 323-334

Thompson, N. (2006). Governing England's National Parks: challenges for the 21st century. *International Journal of Biodiversity Science and Management*, Vol.2, pp. 1 – 3

Toepfer, A. (1957). Wie helfen wir dem Großstadtmenschen? *Die Welt*, 11 may 1957.

[VDN] Verband Deutscher Naturparke. 1995. Die deutschen Naturparke – Aufgaben und Ziele. Luneburg: VDN.

Wascher, D.M. (ed.) (2000). *The face of Europe: policy perspectives for European landscapes*. Tilburg: European Centre for Nature Conservation.

Weber, E. (1976). *Peasants into Frenchmen: The modernization of rural France, 1870-1914*. Stanford/California: Stanford University Press

William Ellis, C. (1947). *On Trust for the Nation*. London: Paul Elek.

11

Comparison of SRTM and ASTER Derived Digital Elevation Models over Two Regions in Ghana – Implications for Hydrological and Environmental Modeling

Gerald Forkuor[1] and Ben Maathuis[2]
[1]*International Water Management Institute,*
[2]*Faculty of Geo-Information Science and Earth Observation, University of Twente,*
[1]*Ghana*
[2]*The Netherlands*

1. Introduction

A Digital Elevation Model (DEM) refers to a quantitative model of a part of the earth's surface in digital form (Burrough and McDonnell, 1998). A DEM consists of either (1) a two-dimensional array of numbers that represents the spatial distribution of elevations on a regular grid; (2) a set of x, y, and z coordinates for an irregular network of points; or (3) contour strings stored in the form of x, y coordinate pairs along each contour line of specified elevation (Walker and Willgoose, 1999). Though there are some disadvantages (Gao, 1997), regular grid DEMs are nowadays the most popular due to their computational efficiency. The use of DEM in this paper, therefore, refers to a regular gridded DEM.

DEMs are useful for many purposes, and are an important precondition for many applications (Kim and Kang, 2001; Vadon, 2003). They are particularly useful in regions that are devoid of detailed topographic maps. DEMs have been found useful in many fields of study such as geomorphometry, as these are primarily related to surface processes such as landslides which can directly be depicted from a DEM (Hengl and Evans, 2009), archaeology as subtle changes due to previous human activity in the sub surface can be inferred on detailed DEMs (Menze *et al.*, 2006), (commercial) forestry, e.g. height of trees and relation to preferred tree stem size (Simard *et al.*, 2006), hydrology, like deriving drainage network and overland flow areas that contribute to (suspended) sediment loads (Lane *et al.*, 1994) and analysis of glaciers (movement of glaciers using multi temporal DEM's) and glaciated terrains (change of glacier thickness by comparison of multi temporal DEM's) (Bishop *et al.*, 2001). Thus for a whole range of different studies, typically topics of interest to geomorphologists, DEMs that provide a good representation of the terrain, are of utmost importance as a starting point for further analysis.

DEMs can be generated using several methods, with varying degrees of accuracy and cost (Flood and Gutelius, 1997). Traditionally, they have been derived from contours that are extracted with photogrammetric techniques from aerial stereo photographs using the parallax displacement on the overlapping images, in conjunction with the flight parameters, to derive elevation information (X,Y,X, Omega, Phi and Kappa) (Petrie and Kennie, 1990). Currently using digital photogrammetric software, when the 6 orientation parameters are recorded during the flight (using an inertial measurement unit (IMU) in conjunction with GPS), direct georeferencing is possible and the initial DEM extraction can be achieved directly. Further post processing might be required though. DEMs are also generated from optical satellite images using similar elevation extraction methods (Jacobson, 2003). In 1986, SPOT was the first satellite to provide stereoscopic images that allowed the extraction of DEMs (Nikolakopoulos *et al.*, 2006). Advances in space technology have now resulted in many more satellites - ASTER, IKONOS, Quickbird and IRS-1C/1D (etc.) - producing stereoscopic images facilitating the extraction of DEMs, both relative, using only orbital information or absolute, integrating also known ground control points. The last procedure is done through provision of actual ground locations in X,Y,Z.

Use of satellite images for DEM generation have a tremendous advantage over traditional methods in that DEMs over large and inaccessible areas can nowadays be easily produced (near) real-time and within a relatively short time and at remarkable cheaper costs. A disadvantage, however, is that optical spectral range requires a cloud free view and appropriate (time of the day) light conditions in order to generate a good quality and high resolution (in accordance with flight parameters) DEM. Radar Interferometry or Interferometric Synthetic Aperture Radar (InSAR) have recently become popular in extracting elevation data. This technique uses two or more Synthetic Aperture Radar (SAR) recordings to generate DEMs, using differences in the phase of the waves returning to the satellite or airborne platform (Rosen *et al.*, 2000). Radars have two main advantages over optical techniques (Massonnet and Feigl, 1998): (1) as an active system, they self-transmit and receive electromagnetic waves. This means image acquisition is independent of natural illumination and therefore images can be taken at night. (2) Observations are not affected by cloud cover since the atmospheric absorption at typical radar wavelengths is very low. National Aeronautics and Space Administrations' (NASA) Shuttle Radar Topographic Mission (SRTM), which produced a near-global DEM, followed this methodology. This system applied a so called single overpass technique using a dual antenna setup. Two systems recorded different wavelength bands (the German Space Research Centre, DLR, operated an X band and NASA a C-band InSAR). For each wavelength, the two antennas' on board of the shuttle were displaced by a certain baseline distance (60 m). When using single overpass, dual antenna techniques, images are recorded at the same time. Using different overpasses from different orbits, recording the same area, mostly the dielectric properties at the Earth surface have changed (given the fact that environmental conditions have changed with respect to a next overpass!) the initial interferogram might show low correlations for locations where environmental conditions have changed in the mean time and therefore the actual elevation might not be properly extracted for these locations.

Another method currently popular in the US and Europe is the Lidar or Airborne Laser Scanning. Due to the fact that these types of recordings are not readily available over Western Africa they are not further discussed here.

Irrespective of the method used, generated DEMs are inevitably subjected to errors, mainly due to the methodology followed (see above) or the various post-processing steps the models have to undergo (e. g. interpolation). It is, therefore, imperative that errors are quantified so as to provide users with first hand information on the accuracy of the DEM.

SRTM and ASTER-derived DEMs are two post-processed elevation datasets that are frequently used for a wide range of applications due to their near-global coverage (Nikolakopoulos and Chrysoulakis, 2006). Dozens of researchers have, over the past few years, carried out a series of global and local assessments of these products. Rodriquez et al. (2006) performed a global validation of the SRTM product using globally distributed set of ground control points derived from Kinematic Global Positioning System (KGPS) transects and National Geospatial-Intelligence Agency's (NGA) level 2 Digital Terrain Elevation Data (DTED). They concluded that SRTM has an absolute height error that exceeds the mission's goal of 16m (90%) often by a factor of two. Hofton et al. (2006) also used data from NASA's Laser Vegetated Imaging Sensor (LVIS) to assess SRTM's accuracy at study sites of variable relief and landcover (i.e. vegetated and non-vegetated terrain). They found out that, under "bare earth" conditions, SRTM data are accurate measurements of LVIS data whereas in vegetated areas, SRTM fell above the ground but below the canopy top, indicating an increase in the vertical offset between SRTM and the LVIS ground data. Jarvis et al. (2004) examined relative and absolute differences between SRTM DEM and a cartographically derived (TOPO) DEM using 1:50,000 scale contours digitized for all Honduras. They validated the two datasets using fifty-nine (59) high accuracy GPS points derived from the National Geodetic Survey (NGS) GPS database from the Central America High Accuracy Reference Network project (NGS, 2003). They also computed slope, aspect, curvature and Pearson's Correlation Coefficient. Their analysis revealed that, SRTM DEM has an average error of 8m as opposed to 20m for the TOPO DEM, although some systematic errors were identified in the SRTM data, related to aspect. They, therefore, concluded that SRTM is more accurate than the 1:50,000 scale cartographically derived DEM for Honduras. Slater et al. (2009) conducted an evaluation of the new ASTER Global Digital Elevation Model (GDEM) using 20 sites in 16 countries. Datasets used for comparison (reference data) with the ASTER GDEM are (1) level 1 and 2 DTED data and (2) control points data (GCPs) photogrametrically derived from satellite imagery. Standard DEM-to-DEM comparisons and DEM-to-control-point comparisons as well as detailed visual analyses of the data were conducted. It was found out that, in almost every area, the mean ASTER GDEM elevations are lower than the reference DEM elevations (biased negatively). The mean ASTER elevations are also lower than the GCPs in every area. Most of the assessments conducted have reported on an absolute accuracy assessment though.

In this study, the C-band SRTM - and ASTER – derived DEMs are compared and validated against a reference DEM for two regions in Ghana. Next to an absolute accuracy assessment also a relative assessment has been conducted. For many environmental and hydrological processes, even if the absolute elevation is not correct, relative to its surrounding regions it might for example exhibit the appropriate flow direction which is the basis of many GIS routines to derive a hydrological network. The reference DEM used was generated using data from a 1:50 000 scale contour map with the TOPOGRID function in the ArcInfo™ software (Hutchinson, 1988, 1989).

1.1 Objectives

In short, the objectives of this work are to:

- Assess how the elevation data from SRTM and ASTER do compare in an absolute manner with respect to a contour based "traditionally derived" elevation model in an absolute comparison. DEMs over different topography and land cover have been selected.

- Derive the "behavior" of the DEMs over different terrain, using multiple tests, to observe their "internal coherence", which is referred to here as "relative assessment". This assessment is especially relevant to geomorphologists, as actual land surface processes do not have a direct bearing on absolute elevation alone, these can also be appropriately studied if relatively to each other the individual elevation grids resemble appropriate elevation change! This topic has hardly been addressed in literature so far!
- Determine implications of the accuracy of the products for hydrological and environmental modeling.

1.2 Study sites

The study is conducted for two sites in Ghana (Figure 1). These sites fall in different agro climatic zones, have different elevation ranges and differ in landcover. Site 1, which climatologically falls in the Guinea Savannah zone (northern Ghana), has an elevation range of about 400 m. It is fairly flat, with an average slope of 0.9^0. The major landcover types are deciduous woodland (55 %) and shrubland (36 %). Site 2 lies between two climatic zones – moist semi-deciduous forest and transitional zones. It has an elevation range of about 780 m, with an average slope of 3.3^0. Though also fairly flat, this site has a range of mountains that borders the Volta Lake. The dominant landcover types are forest (52 %) and woodland/shrubland (34 %).

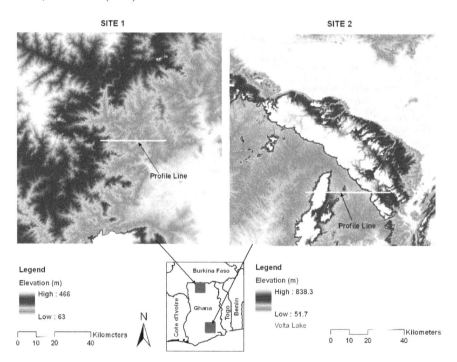

Fig. 1. Map of the study sites

2. Data

2.1 Aster GDEM

The Advanced Spaceborne Thermal Emission and Reflection Radiometer (ASTER) is an advanced multispectral imager that was launched onboard NASA's Terra spacecraft in December, 1999. ASTER has three spectral bands in the visible near-infrared (VNIR), six bands in the shortwave infrared (SWIR), and five bands in the thermal infrared (TIR) regions, with 15-, 30-, and 90-m ground resolution, respectively (Yamaguchi *et al.* 1998). The VNIR subsystem has one backward-viewing (27.7^0 off-nadir) instrument for stereoscopic observation in the along-track direction, making imagery acquired by this satellite suitable for DEM generation. Other important properties that make ASTER data suitable for DEM generation are the platform altitude (705 km) and its base-to-height ratio of 0.6 (Abrams 2000; Hirano *et al.* 2003).

The Ministry of Economy, Trade and Industry of Japan (METI) and the United States NASA recently released a global DEM (ASTER GDEM) derived from ASTER images acquired since its launch (1999) to the end of August, 2008. It covers land surfaces between 83^0 N and 83^0 S, comprising of 22 600 1^0 –by- 1^0 tiles. The GDEM is provided at a one arcsec resolution (30 m) and referenced to World Geodetic System (WGS) 1984. Elevations are computed with respect to the WGS 84 EGM96 geoid. The vertical accuracy of the DEM data generated from the Level-1A data is 20 m with 95% confidence without ground control point (GCP) correction for individual scenes (Fujisada *et al.*, 2005).

METI and NASA acknowledges that Version 1 of the ASTER GDEM is "research grade" due to the presence of certain residual anomalies and artifacts in the data that may affect the accuracy of the product and hinder its effective utilization for certain applications. For this study, two 1^0 –by- 1^0 tiles in Ghana (see figure 1) were download from NASA's Warehouse Inventory Search Tool (WIST - https://wist.echo.nasa.gov/api/)

2.2 SRTM DEM

The SRTM (Werner, 2001; Rosen *et al.*, 2001a), undertaken by NASA and the NGA, collected interferometric radar data which has been used by the Jet Propulsion Laboratory (JPL) to generate a near-global (80% of earth's land mass) DEM for latitudes smaller than 60^0. SRTM has been the first mission using space-borne interferometric SAR (InSAR). The SRTM mission has been a breakthrough in remote sensing of topography (van Zyl, 2001), producing the most complete, highest resolution DEM of the world (Farr *et al.*, 2007). An extensive global assessment revealed that the data meets and exceeds the mission's 16m (90 percent) absolute height accuracy, often by a factor of two (Rodríguez *et al.*, 2006). Since its release in 2005, the user community has embraced the availability of SRTM data, using the data in many operational and research settings.

SRTM data for this study was downloaded from the website of the Consultative Group on International Agricultural Research Consortium for Spatial Information (CGIAR-CSI - http://srtm.csi.cgiar.org). Data available from this site has been upgraded to version 4, which was derived using new interpolation algorithms and better auxiliary DEMs. This version, thus, represent a significant improvement from previous ones.

2.3 Reference DEM

The reference DEM was generated from the hypsographic (contours) and hydrographic (rivers/streams) data in the 1:50,000 topographical map series produced by the Survey Department of Ghana (SDG). The contours used have an interval and vertical accuracy of 50 feet (≈ 15m). This accuracy supersedes that of the 16m and 20m of SRTM and ASTER respectively, hence its use as a reference. Map sheets covering the study sites were merged prior to generating the DEM.

The SDG, in January 2009, adopted the WGS 1984 (UTM) coordinate system (Daily Graphic, 14th October 2008). Prior to this, the department's maps (including the ones used in this study) were produced based on a local datum and the "war office" ellipsoid. Thus, there was the need to transform the data from the old system (Ghana grid) into the new adopted system (WGS 84 UTM). Table 1 shows the transformation parameters, as published by the SDG, used for the conversion.

The TOPOGRID function in the ArcInfo™ software was used to interpolate the transformed contour data into a DEM. There are many interpolation methods used to generate a DEM from contour data (Hengl and Evans, 2009). Comparison of these techniques has been extensively carried out by many researchers (Wise, 2000). Li et al. (2005) points out that there is no universal interpolation technique that is clearly superior, and appropriate for all sampling techniques and DEM applications. Despite this conclusion, Hengl and Evans (2009) recommend that, where possible, algorithms that can incorporate secondary information (such as layers representing pits, streams, ridges, scarps or break lines) should be implemented for DEM interpolation. One such method, and the one used in this study, is the ANUDEM algorithm by Hutchinson (1988, 1989), and implemented as the TOPOGRID function in the ArcInfo™ software. ANUDEM is based on the discretized thin-plate spline technique (Wahba, 1990), and is an iterative DEM generation algorithm that produces hydrologically correct DEMs. The algorithm starts with a coarse grid, and then enforces drainage conditions, the spatial resolution is increased, and then drainage enforcement is performed again, and so on, until the desired resolution is reached (Hengl and Evans, 2009).

Local Geodetic Datum		Transformation Parameters										
Name	Code	Pub. Date	ΔX	ΔY	ΔZ	Rx (Rads)	Ry (Rads)	Rz (Rads)	Scale factor	Xo	Yo	Zo
Ghana	GHA	2008	-196.58	33.383	322.552	0	0	0	1	0	0	0

Source: Daily Graphic of Ghana, 14th October 2008

Table 1. Transformation parameters (War Office to WGS 84)

The grid resolution chosen for the reference DEM was based on equation (1). In cases where a DEM is based on contours, a suitable grid resolution can be estimated from the total length of the contours (Hengl and Evans, 2009). It is important to choose a grid resolution that optimally reflect the complexity of the terrain and represent the majority of the geomorphic features (Smith et al., 2006).

$$\Delta S = \frac{A}{2.\sum L} \tag{1}$$

where "A" is the area of the study site (km²) and "L" is the cumulative length of all contour
lines (km).

Table 2 shows the implementation of equation (1) for both study sites and the resultant grid
resolution. Based on these results, a cell size of 90 m was chosen in generating the reference
DEM. The choice of a 90 m spatial resolution reflects the complexity of terrains for both
study sites and allows direct comparison with the 90 m SRTM data.

Site	Area (Km²)	Length of contours (km)	Grid resolution (m)
1	12100	21736.0	278.3
2	12100	61875.6	97.8

Table 2. Suitable grid resolution for the Reference DEM

3. Methodology

3.1 Data preparation

DEMs of the study sites were transformed into the same projection system – Universal
Transverse Mercator (UTM) zone 30 north. WGS 1984 was selected as both datum and
spheroid. Part of the Volta Lake falls within site 2. For this reason, a mask was prepared and
used to mask out the lake area on all DEMs (ASTER, SRTM and Reference). The original 30
m resolution of the ASTER GDEM was resampled to 90 m to enable comparison with the
other DEMs. After resampling, a low pass 3 x 3 filter was applied to all DEMs to remove
possible outliers still remaining in the data. In practice, smooth models of topography and a
small amount of smoothing of DEMs prior to geomorphometric analysis have proved more
popular among geomorphometricians, although no single smoothing approach is absolutely
superior for all datasets and study areas (Hengl and Evans, 2009). No misalignment between
DEMs was observed, thus co-registration was not necessary. Table 3 presents summary
statistics of the DEMs compared for both sites.

3.2 Comparison of DEMs

Two main approaches were used to compare and validate the elevation products against the
reference. These are: (1) determining the accuracy of the *elevation values* of the products
(absolute accuracy) and (2) determining the accuracy of terrain derivatives of the products
(relative accuracy).

3.2.1 Accuracy of elevation values

This was achieved by performing DEM differencing, profiling and correlation plots.

- **DEM differencing**: This was performed to derive elevation error maps. Root mean
  square error (RMSE), a common measure of quantifying vertical accuracy in DEMs, was

calculated for each error map. In addition, skewness and kurtosis (King and Julstrom, 1982) was calculated for each error map. Skewness is a unitless measure of asymmetry in a distribution (Shaw and Wheeler, 1985). Negative skewness indicates a longer tail to the left, while positive skewness indicates a longer tail to the right. Excess kurtosis is a unitless measure of how sharp the data peak is. A value larger than zero (0) indicates a peaked distribution, while a value less than zero (0) indicates a flat distribution. Percentage of pixels falling within different error ranges was also determined.

- **Profiling**: Horizontal profiles were created on the DEMs and compared. Profile lengths were 35 km and 45 km for site 1 and 2 respectively. A graph of elevation against distance was produced for comparison. Figure 3 (*a* and *b*) shows profile graphs for sites 1 and 2.

- **Correlation scatter plots**: This was performed to assess the level of correlation between the DEMs. It was difficult making a scatter plot from all the pixels in a DEM, as each DEM contained over a million pixels. For this reason, the scatter plots are based on randomly selected points/pixels. In all, about 6000 points/pixels were randomly selected from each DEM. They are an aggregation of pixels randomly selected from different landcover types identified in each site. For each site, two scatter plots were produced and correlation coefficient determined.

Dataset	Description	Min	Max	Mean	Standard deviation	Skewness	Kurtosis
				Site 1			
Reference	Elevation	105	463	213.0	63.2	0.45	2.3
	Slope	0	30.2	0.86	0.94	9.5	183.6
SRTM	Elevation	111	467	216.9	63.5	0.45	2.4
	Slope	0	31.9	0.97	0.9	11.1	240.6
Aster	Elevation	77	464	209.87	64.8	0.4	2.3
	Slope	0	29.9	1.06	0.9	9.3	182.2
				Site 2			
Reference	Elevation	44.8	836.0	238.9	139.8	1.2481	3.8751
	Slope	0	45.1	3.3	4.2	2.523	10.758
SRTM	Elevation	51.7	838.3	241.0	139.7	1.2499	3.9459
	Slope	0	47.1	3.5	4.1	2.5949	11.311
Aster	Elevation	41.3	848.7	235.6	139.6	1.2479	3.9275
	Slope	0	44.3	3.7	4.0	2.4926	10.561

Table 3. Summary statistics of DEMs analysed for both sites

3.2.2 Accuracy of terrain derivatives

In this assessment, the three DEMs were first preprocessed to obtain a hydrologically consistent elevation model (filling local depressions). Flow direction, using a Deterministic-8

Comparison of SRTM and ASTER Derived Digital Elevation Models over Two Regions in Ghana – Implications for Hydrological and Environmental Modeling

267

flow direction algorithm, flow accumulation and drainage maps were subsequently generated. Next, a common outlet location was used to extract an upstream catchment area from the DEMs and the attributes of the derived catchments compared. This analysis was conducted for site 1 alone due to two main reasons: (1) the location of the Volta Lake within site 2 would result in a large number of relatively small catchments that directly drain into the lake, and (2) the relief of site 1 is relatively gentle (average slope = 0.9^0) and for the accuracy of terrain derivatives this is critical since in steeply sloping terrain the flow direction is assigned correctly. Using statistics extracted, the following analyses and comparisons were performed.

- **Longitudinal Profiles:** these were extracted at 500m interval, along the longest flow path, with the respective depression- filled DEMs as source. Profiles of the three DEMs were plotted and compared.
- **Horton Plot:** In hydrology, the geomorphology of the watershed, or quantitative study of the surface landform, is used to arrive at measures of geometric similarity among watersheds, especially among their stream network. The quantitative study of stream networks was originated by Horton. Horton's original stream ordering was slightly modified by Strahler, and Schumm added the law of stream areas. Number of streams of successive order, the average stream length of successive order and the average catchment area of successive order is found to be relatively constant from one order to another. Graphically this can be visualized by construction of a Horton plot. The Horton plots show the relationship between Strahler order and total number of Strahler order stream segments for a given order, average length per Strahler order and average catchment area per Strahler order, as well as the bifurcation (Rb), channel length (Rl) and stream area ratio's (Ra), by means of a least square regression line. Using the extracted drainage data from the three DEMs, Horton's statistics were derived and the results graphically displayed by plotting Strahler order on the X-axis and the number of drainage channels, stream length and stream area on a log transformed Y axis.
- **Geomorphological Instantaneous Unit Hydrograph (GIUH):** The GIUH model is based on the theory proposed by Rodriquez-Iturbe and Valdes and its subsequent generalization by Gupta. According to the theory, the unit input (unit depth of rainfall) is considered to be composed of an infinite number of small, non-interacting drops of uniform size, falling instantaneously over the entire region. The basin geomorphology plays an important role in the transition of water from the overland region to channels (streams) and also from the channels of one order to the other (Bhadra *et al.*, 2008). Using extracted statistics such as Horton's ratios, length of highest order stream, maximum order, etc., a GIUH was calculated, assuming identical effective precipitation and effective holding capacity for each of the datasets. No infiltration was assumed. A user-friendly event based computerized GIUH model, "GIUH_CAL" (Bhadra *et al.*, 2008) was applied to derive the direct runoff hydrograph of an assumed uniform storm event for all three DEMs and compared.
- **Topographic Index (TI):** TI is a topography-based concept for watershed hydrology modeling which has been widely used to study the effects of topography on hydraulic processes (Wolock and McCabe, 1995). The TI, $ln(a/tan\ b)$, is the natural logarithm of the ratio of the specific flow accumulation area "a" to the ground surface slope "$tan\ b$". Surface slope can be evaluated from DEMs. The specific flow accumulation area is the total flow accumulation area (or upslope area) "A" through a unit contour length "L".

When using a DEM, L can be defined as being equal to the horizontal resolution of the DEM (Pan *et al.*, 2004). In this study, the algorithm applied uses a Deterministic-8 flow direction algorithm to obtain the (specific) flow accumulation area. The continuous Topographic Index data range obtained for each of the DEMs was classified into integer classes and plotted for comparison purposes.

4. Results and discussion

4.1 Accuracy of elevation values

Table 4 presents the statistics of the error maps obtained for both sites, whereas figure 2 (*a* - *d*) shows the spatial distribution of the errors and the percentage of pixels that falls in different error ranges. It is quite evident that better results were obtained for site 1 than site 2. This is manifested in the RMSEs obtained for site 1. Although all RMSEs fall within predefined vertical accuracy specification (Slater *et al.*, 2006; Fujisada *et al.*, 2005), results for site 1 indicate that the products (of site 1) are three-times better than that of site 2. This could be due to the physical characteristics of site 1, which has a relatively flat terrain with a mean slope of about 0.9⁰. It can, thus, be said that both products do better on flat and less complex terrain than would do on hilly and mountainous terrain as in site 2.

Difference Map	Min	Max	Mean	Standard deviation	RMSE	Skewness	Kurtosis
Site 1							
ASTER - *Reference*	-82.92	118.01	-3.30	6.05	5.46	1.4326	20.884
SRTM - *Reference*	-70.96	143.57	3.67	5.70	4.95	2.3606	43.820
Site 2							
ASTER - *Reference*	-245.56	233.78	-3.323	20.41	18.76	-0.4743	10.512
SRTM - *Reference*	-229.89	204.44	2.089	16.08	14.54	-0.7992	15.054

Table 4. Difference statistics of the study sites

Table 4 further reveals that, compared to the reference DEM, SRTM has a better vertical accuracy than the ASTER GDEM. In both sites, a smaller RMSE was obtained for SRTM than ASTER GDEM. This finding is in line with the pre-launch vertical accuracy of 16m for SRTM (Hensley *et al.*, 2001) and 20m for ASTER GDEM (Fujisada *et al.*, 2005; Slater *et al.*, 2009).

Figures 2*a* and 2*b* shows the spatial distribution of errors in the ASTER GDEM for both sites. The graphs and statistics indicate that ASTER GDEM elevations are generally lower, compared to the reference DEM (i. e biased negatively). In other words, the ASTER GDEM underestimates elevation on both sites. Statistics from site 1 indicate that 74% of pixels fall below zero (0), whereas 58% of pixels fall below zero (0) in site 2. This further reveals that, though ASTER GDEM generally underestimates elevation, this underestimation is more pronounced on flat and less complex terrains (as in site 1) than in hilly and complex terrains.

Figures 2*c* and 2*d* shows that SRTM have the directly opposite characteristic – overestimates elevation. Elevation differences are positively biased, resulting in majority of pixels being greater than zero (0). Statistics from site 1 indicate that about 78% of pixels were greater than zero (0), whereas about 60% was greater than zero (0) in site 2. This overestimation may be partly due to the fact that SRTM records the reflective surface and, thus, may be positively biased with respect to the bare earth when foliage is present. This under- and overestimation of ASTER GDEM and SRTM respectively has been noted in previous studies (Slater *et al.*, 2009).

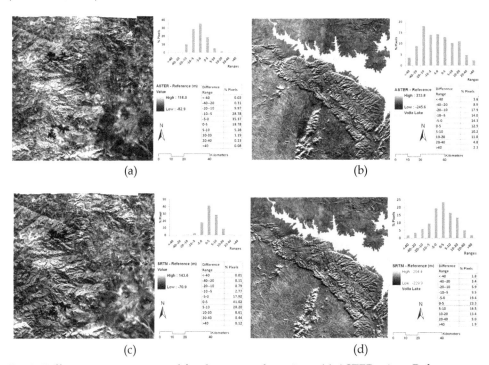

Fig. 2. Difference maps computed for the two study regions. (a) ASTER minus Reference (Site 1). (b) ASTER minus Reference (Site 2). (c) SRTM minus Reference (Site 1). (d) SRTM minus Reference (Site 2).

Apart from generating error maps, horizontal profiles were created on the DEMs using the 3-D analyst extension in ArcGIS®, and the data exported to excel for comparison. Figure 1 above shows the location of the profile lines, whereas figure 3 (*a* and *b*) below shows the comparison between the three (3) DEMs for both sites. The results obtained in this section further confirm the earlier finding that ASTER GDEM underestimates elevation whereas SRTM overestimates. Figure 3*a* clearly shows how bad ASTER GDEM performs on lowlands – its profile line is consistently below that of SRTM and the reference DEM. A visual inspection of figure 3 reveals that, the magnitude of overestimation of SRTM is less than the magnitude of underestimation of ASTER GDEM. In other words, SRTM is "closer" to the reference than ASTER GDEM. This further confirms that SRTM has an accuracy superior to ASTER.

Figure 4 (*a* – *d*) shows the correlation plots obtained for both sites. As stated earlier, these plots are based on a random selection of points representing different land covers. Results from site 1 indicate that both products have the same correlation coefficient with the reference DEM, which is slightly different from earlier results discussed above. This could be due to the number and distribution of points selected (i. e. 6000 out of over a million points). The plot for site 2, however, indicates that SRTM is slightly better correlated to the reference than ASTER GDEM.

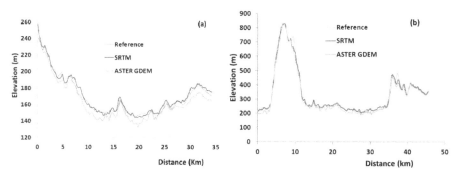

Fig. 3 (a). Comparison of profile lines derived from all DEMs for site 1. (b). Comparison of profile lines for site 2

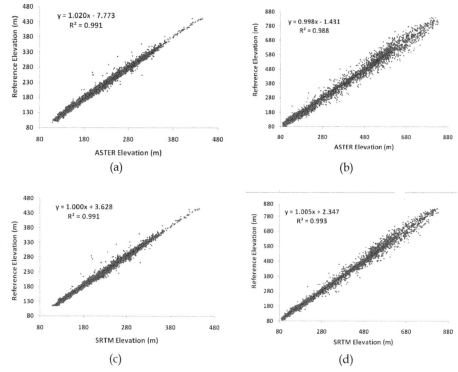

Fig. 4. Correlation plots for the two study sites. (a) ASTER versus Reference (Site 1). (b) ASTER versus Reference (Site 2). (c) SRTM versus Reference (Site 1). (d) SRTM versus Reference (Site 2).

Comparison of SRTM and ASTER Derived Digital Elevation Models over Two Regions in Ghana – Implications
for Hydrological and Environmental Modeling
271

4.2 Accuracy of terrain derivatives

Results of the hydro-processing to extract catchments and drainage information, using a common outlet, from the three DEMs are shown in Table 5. The upstream catchment area extracted for a common outlet location is largest for the Reference DEM, the catchment areas of ASTER and SRTM are -1% and -0.7% respectively. Although the same flow accumulation threshold map has been used for all DEMs, larger differences appeared during the drainage extraction process. This resulted in the largest deviations with respect to the Reference DEM of the drainage length and drainage density for the SRTM DEM, +7.6% and +4.5% respectively.

Catchment characteristics	Reference	ASTER	SRTM
Area (km2)	7,189.97	7,118.36	7,140.06
Perimeter (m)	442,767.37	475,829.21	463,395.49
Total Drainage Length (m)	2,725,840.8	2,890,504	2,911,511.2
Drainage Density (m/km2)	379.12	406.06	407.77
Longest Flow Path Length (m)	189,275.21	194,572.84	197,697.76
Longest Drainage Length (m)	186,556	192,018.2	194,910.4
Sinuosity	1.787	1.816	1.864

Table 5. Attribute values for extracted catchments and longest flow path

Figure 5 shows the longitudinal profiles extracted at 500 m interval along the longest flow path, with the various depression-filled DEMs as the elevation source. The graph confirms results of the absolute accuracy assessment, i.e. ASTER's underestimation and SRTM's overestimation of elevation.

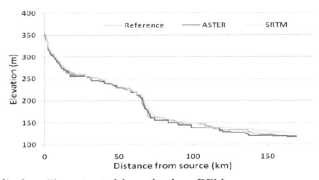

Fig. 5. Longitudinal profiles extracted from the three DEMs

Horton statistics were computed using the extracted drainage data from the three DEMs. The statistical values are shown in Table 6 while Figure 6 shows the resulting Horton plot for Strahler stream orders 1 to 5. The Strahler order is plotted on the "X" axis while the number of drainage channels, stream length and stream area are plotted on a log transformed "Y" axis. According to Horton's law the values obtained should plot along a straight line and this can be used as an indicator that the parameters used for drainage extraction are properly selected (Chow *et al.*, 1988).

SRTM and ASTER have a larger number of drainage lines per Strahler order, especially for the lower order streams and therefore the stream length per order is less for SRTM and ASTER compared to the Reference DEM. The stream area shows a similar tendency. The stream lengths for the fifth and sixth order are the main deviating phenomena; smaller and larger for the Reference DEM, compared to ASTER and SRTM respectively. The Horton ratio's, calculated excluding the lowest and highest stream orders, shows that SRTM has slightly higher ratio values compared to those derived from the Reference DEM for all ratio's. ASTER derived ratio's for "Rb" and "Rl" are in between, the area ratio of ASTER is strongly deviating, only 3.99. In general the trend of the ratios as plotted in Figure 6 show a similar geomorphological structure for ASTER and SRTM and only deviates notably with respect to the Reference DEM for the Length Ratio. The "Rb", "Rl" and "Ra" values vary normally between 3 and 5 for "Rb", between 1.5 and 3.5 for "Rl" and between 3 and 6 for "Ra" (Rodrıguez-Iturbe, 1993). For all DEMs the derived ratios are within the ranges given.

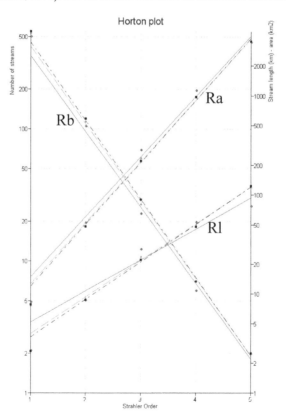

Fig. 6. Horton plot (Legend: Red = Reference; Blue = ASTER; Black = SRTM)

Order	Reference			ASTER			SRTM		
	No. of streams	Length (km)	Area (km2)	No. of streams	Length (km)	Area (km2)	No. of streams	Length (km)	Area (km2)
1	502	2.7	8.34	537	2.66	7.72	550	2.63	7.84
2	105	8.66	52.55	114	8.93	48.83	119	8.67	47.59
3	23	28.2	284.62	27	23.79	237.75	29	22.03	221.73
4	6	53.7	1146.61	7	49.48	972.54	7	47.88	975.53
5	2	77.42	3496.54	2	120.9	3540.11	2	123.14	3549.87
6	1	172.09	7190.37	1	145.98	7138.2	1	148.56	7154.38
Horton Ratio's (calculated excluding lowest and highest stream order)									
	Ratio Reference		Ratio ASTER		Ratio SRTM				
Rb	3.55		3.61		3.64				
Rl	2.22		2.26		2.28				
Ra	4.55		3.99		4.6				

Table 6. Horton statistics

4.3 Implications for hydrological and environmental modeling

The reported accuracies and geomorphological behavior of the studied DEMs have numerous implications for hydrological and environmental modeling in the study regions. Topography is a crucial land surface characteristic and controls many earth processes (Hutchinson, 1996). For this reason, topography needs to be adequately represented (as in the form of a DEM) to ensure that modeling results, e.g. predicting surface/ sub-surface runoff, erosion estimates, etc. are as much as possible close to observed values. The implications of using inappropriate DEMs in hydrological and environmental modeling have been amply studied and reported in the literature. For example, Datta and Schack-Kirchner (2010), in reviewing erosion studies conducted in the Indian lesser Himalayas, attributed the large variations in the range of soil loss estimates chiefly to poor description of the terrain in the form of a DEM. They compared erosion relevant topographical parameters – elevation, slope, aspect and LS factor – derived from DEMs of different source and accuracy and concluded that the choice of a DEM for soil erosion modeling has a significant impact on the relevant topographical parameters and, consequently, on the modeling results. In a related study, Rojas et al. (2008) studied the effects of DEM grid sizes – 30m, 90, 150, 210, 270 and 330 - on results of modeling upland erosion and sediment yield from natural watersheds. They found that using different resolution DEMs (30m – 330m) significantly reduces land surface slopes and channel network topology, resulting in varied upland erosion estimates. Walker and Wildgoose (1999) studied the implications of using different source DEMs – cartometric, photogrammetric and ground truth - of varying resolution and accuracy on key hydrologic statistics. They found catchment sizes and stream networks derived from the cartometric and photogrammetric DEMs to be significantly different from that of the ground truth. This was particularly the case in smaller catchments where a localized error in elevation may direct a major stream line in a wrong direction. They concluded that the use of published DEMs for determining catchment boundaries and stream networks must be done with care by comparing results with that of a ground truth DEM or a higher accuracy DEM.

In light of the above, this study computed two indices - GIUH and TI – to assess the suitability of using the published elevation products under consideration in hydrological and environmental modeling. A GIUH enables the determination of the hydrological response of a catchment to rainfall by taking into consideration its geomorphology. The runoff volumes generated at the outlet of a catchment and time-to-peak are dependent on the topography of the overland regions as well as the transmission surfaces. The reliability of these parameters (runoff & time-to-peak) is, therefore, dependent on how well a catchment's terrain (landform) is represented by the underlying elevation product (DEM). A GIUH was calculated using the three DEMs and their responses compared. Apart from changing the Horton statistics for the respective DEMs, all other variables, such as the rainfall intensity and a mean holding time of 5 hours, remained constant. Result of the GIUH analysis is shown in Figure 7. The figure shows that the direct runoff hydrographs for the Reference DEM and SRTM DEM are comparable with respect to the volume as well as the timing (i.e. time to peak). The ASTER DEM, however, shows a delay in the rising limb and a higher peak discharge. Considering that Horton statistics are the only variable in the analysis (with all others remaining constant), the behavior of ASTER can be attributed to its representation of the catchment's geomorphology. Although ratios obtained in the Horton analysis fall within acceptable ranges, the stream area ratio (Ra) obtained for the ASTER DEM (3.99), which strongly deviates from that of the other two DEMs, is believed to have caused the noted delay in rising limb and higher peak discharge.

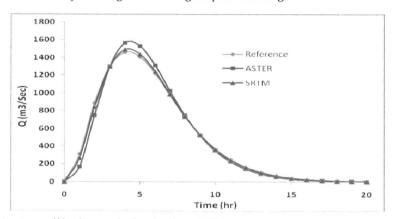

Fig. 7. Direct runoff hydrographs for the three DEMs

Figure 8 shows results of the TI analysis. Continuous Topographic Index data range obtained for each of the DEMs was classified into integer classes, and the percentage of pixels falling in each was determined and plotted. The graph shows a notable difference between the reference DEM and the two global DEMs. In order to attribute reasons for the results obtained, slope maps of the three DEMs were created and analyzed. Slope was chosen and analyzed due to the fact that TI is a function of the local slope angle acting on the pixel. The analysis revealed that, the distribution of slopes was quite different in all three DEMs. The reference DEM was found to have about 67% of its area having slopes up to 1^0, while for the ASTER and the SRTM DEMs, this slope class accounted for nearly 50% and slightly over 50% respectively. In the case of SRTM the radar reflective surface seems

to cause a more "rough" topography and the difference with respect to ASTER may be
attributed to the resampling of the original 30m to 90m for purposes of comparison with
the other DEMs. Previous studies, e.g. Rojas et al. (2008), have noted that resampling
DEMs to coarser resolutions generally reduces surface slopes and changes channel
topology. Slopes have also been found to be highly sensitive to varying DEM resolution
and accuracy (Datta and Schack-Kirchner, 2010). Zhang and Montgomery (1994), in their
investigation of the effect of grid size on TI concluded that increasing DEM grid sizes
results in increased mean TI due to increased contributing area and decreased slopes. This
fact was also emphasized by Wilson et al. (2000). A reasonable conclusion can, thus, be
drawn that the low percentage of pixels up to 1^0 in ASTER, and the resultant TI curve,
could be attributed to the resampling of the DEM from the original 30m to 90m. The
similarity between the ASTER and SRTM curves indicate that SRTM failed to produce the
desired slope compared to the reference.

Results obtained in the GIUH and TI analysis is a clear demonstration of the effects of
topography on hydrological processes and the need to select the right DEM (resolution and
accuracy) for hydrological and environmental modeling. Use of a DEM that does not
adequately represent the surface landforms of a catchment has the tendency of producing
erroneous modeling results. With the exception of the TI, the comparison in this study
consistently revealed SRTM to have a superior accuracy compared to ASTER DEM,
although its resolution is coarser. This is in line with results of previous studies (Datta and
Schack-Kirchner (2010).

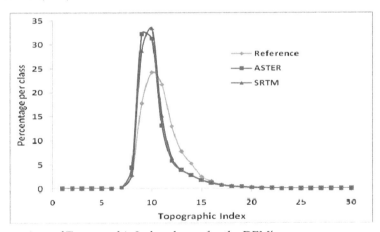

Fig. 8. Comparison of Topographic Index classes for the DEM's

5. Conclusions

In this study, two near-global DEMs - SRTM and ASTER – are compared and validated
against a reference DEM for two sites in Ghana. The reference DEM used was generated
using hypsographic and hydrographic data from a 1:50 000 topographical map produced by
the SDG. DEM differencing, profiling, correlation plots, extraction of catchment area and
drainage network, computation of Horton statistics and GIUH are some of the methods
employed in the comparison.

Results obtained indicate that, for the two sites selected, both SRTM and ASTER GDEM meet their predefined vertical accuracy specifications of 16m and 20m respectively. This is in line with results of previous studies (Fujisada *et al.*, 2005; Rodriquez *et al.*, 2006). It was realized that SRTM has a higher vertical accuracy (in terms of RMSE) than ASTER GDEM for both sites. RMSEs ranged between 4.9 and 5.5 (site 1) and 14.5 and 18.8 (site 2) for SRTM and ASTER respectively. The vertical accuracy of both products, thus, increases (by a factor of 3) on flat and less complex terrain (i.e. site 1). Analyses conducted revealed that ASTER GDEM underestimates elevation (i.e. negatively biased), even if fill routines are applied. SRTM, on the other hand, overestimates elevation, which may be partly due to the fact that SRTM records the reflective surface and, thus, may be positively biased with respect to the bare earth when foliage is present. The underestimation of ASTER is more pronounced on flat and less complex terrain (site 1), and of a greater magnitude than the overestimation of SRTM. Results of horizontal profiling on site 1 showed that the elevation of ASTER GDEM is consistently lower than that of the other two. In areas that are heavily vegetated, the effect of the under- and overestimation of ASTER and SRTM respectively can be reduced by constructing an average DEM (i.e. ASTER+SRTM/2), which will have an absolute accuracy between that specified for the two global DEMs.

In the relative accuracy assessment, the Horton plot revealed that SRTM and ASTER DEMs have a similar geomorphological structure as compared to the Reference DEM. The main deviation is with respect to the stream area ratio (Ra) of ASTER. All other ratio values extracted are within the ranges as given for the various ratio's (Rodriguez-Iturbe, 1993). The calculated direct runoff hydrograph showed similar response for the Reference and SRTM DEMs. ASTER DEM, however, showed a delayed rising limb and a higher peak discharge. This might be due to the deviating area ratio obtained for ASTER DEM. The calculated Topographic Index for the DEMs showed a substantial difference between the global DEMs and the Reference DEM. This difference is to be attributed to the gentle slopes that prevail in the area (site 1) analyzed.

In summary, the study has revealed that SRTM is "closer" to the Reference DEM than ASTER, although both products are useful and are an excellent replacement for local 1:50 000 hypsographic data both in absolute and relative terms. The relative assessment further confirms that various surface processes can be appropriately studied when using these global elevation data sets, which is a great asset to geomorphologists. Here the relative assessment conducted is more focused to hydrological processes, one of the terrain processes important in geomomorpholgy.

6. References

Abrams, M., 2000, ASTER: data products for the high spatial resolution imager on NASA's EOS-AM1 platform. *International Journal of Remote Sensing*, 21, pp. 847–861.

Bhadra, A., Panigrahy, N., Singh, R., Raghuwanshi, N. S., Mal, B. C. and Tripathi, M. P., 2008, Development of a geomorphological instantaneous unit hydrograph model for scantily gauged watersheds. Environmental Modelling & Software, 23, pp. 1013-1025.

Bishop, M.P., Bonk, R., Kamp, U. and Shroder, J.F., 2001, Topographic analysis and modeling for alpine glacier mapping. *Polar Geography*, 25, pp. 182-201

Burrough, P. A. and McDonnell, R. A., 1998, Principles of geographic information systems, 333 pages (New York: Oxford University Press)

Chow V. T., Maidment, D. R. and Mays, L. W., 1988, *Applied Hydrology*, pp. 166-170 (Singapore: McGraw-Hill)

Datta, P. S. and Schack-Kirchner, H, 2010, Erosion Relevant Topographical Parameters Derived from Different DEMs – A Comparative Study from the Indian Lesser Himalayas. *Remote Sensing*, 2, pp. 1941-1961.

Farr, T. G. et al., 2007, The Shuttle Radar Topography Mission. *Reviews of Geophysics*, 45, 33 pages.

Flood, M. and Gutelius, B., 1997, Commercial Implications to Topographic Terrain Mapping Using Scanning Airborne Laser Radar. *Photogrammetric Engineering and Remote Sensing*, 63, pp. 327.

Fujisada, H., Bailey, G. B., Kelly, G. G., Hara, S. and Abrams, M. J., 2005, ASTER DEM Performance. *IEEE transactions on Geoscience and Remote Sensing*, 43, pp. 2707-2714

Gao, J., 1997, Resolution and terrain representation by grid DEMs at a micro-scale. *International Journal of Geographical Information Science*, 11, pp. 199-212.

Hengl, T. and Evans, I. S., 2009, Mathematical and Digital models of the Land Surface. In *Geomorphometry: concepts, software, applications*, T. Hengl, and H. I. Reuter (eds), pp. 31-63 (The Netherlands: Elsevier)

Hensley, S., Munjy, R. and Rosen, P., 2001, In Interferometric Synthetic Aperture Radar (IFSAR), Digital Elevation Model Technologies and Applications: The DEM Users Manual, D. Maune (Ed.), pp. 143–206 (Bethesda, MD: American Society for Photogrammetry and Remote Sensing).

Hirano, A., Welch, R. And Lang, H., 2003, Mapping from ASTER stereo image data: DEM validation and accuracy assessment. *Journal of Photogrammetry and Remote Sensing*, 57, pp. 356-370.

Hofton, M., Dubayah, R., Blair, J. B. and Rabine, D., 2006, Validation of SRTM elevations over vegetated and non-vegetated terrain using medium footprint Lidar. *Photogrammetric Engineering and Remote Sensing*, 72, pp. 279-285

Hutchinson, M. F., 1988, Calculation of hydrologically sound digital elevation models. In *Proceedings of the Third International Symposium on Spatial Data Handling*, Sydney, Australia.

Hutchinson, M. F., 1989, A new procedure for gridding elevation and stream line data with automatic removal of spurious pits. *Journal of hydrology*, 106, pp. 211-232

Hutchinson, M. F, 1996, A locally adaptive approach to the interpolation of digital elevation models. The Proceedings of Third International Conference/Workshop on Integrating GIS and Environmental Modeling, 21–26 January 1996, Santa Fe, NM.

Jacobson, K., 2003, DEM Generation from Satellite data. Available online at: http://www.earsel.org/tutorials/Jac_03DEMGhent_red.pdf (Accessed 1st July 2011)

Jarvis, A., Rubiano, J., Nelson, A., Farrow, A. and Mulligan, M, 2004, Practical use of SRTM data in the tropics – Comparisons with digital elevation models generated from cartographic data. Working Document No. 198, 32p (Cali, Columbia: CIAT)

Kim, S. and Kang, S., 2001, Automatic Generation of a SPOT DEM: Towards Coastal Disaster Monitoring. *Korean Journal of Remote Sensing*, 17, pp. 121-129

King, R. S. and Julstrom, B., 1982, Applied Statistics Using the Computer (Sherman Oaks, CA: Alfred Pub. Co.)

Lane, S. N., Chandler, J. H. and Richards, K. S., 1994, Developments in monitoring and modelling small-scale river bed topography. *Earth Surface Processes Landforms*, 19, pp. 349–368.

Li, Z., Zhu, Q. and Gold, C., 2005, Digital Terrain Modeling: Principles and Methodology, 319 pages (Florida: CRC Press)

Massonnet, D. and Feigl, K. L., 1998, "Radar interferometry and its application to changes in the earth's surface". *Reviews of Geophysics*, 36, pp. 441–500.

Menze, B. H., Ur, J. A. and Sherratt, A. G., 2006, Detection of Ancient Settlement Mounds: Archaeological Survey Based on the SRTM Terrain Model. *Photogrammetric Engineering & Remote Sensing*, 72, pp. 321-327

NGS (National Geodetic Survey), 2003, Central American HARN, 2001 GPS Survey Project. Available at:
http://www.ngs.noaa.gov/PROJECTS/Mitch/Honduras/stationlist.htm

Nikolakopoulos, K. G. and Chrysoulakis, N., 2006, Updating the 1:50.000 topographic maps using ASTER and SRTM DEM: the case of Athens, Greece. Remote Sensing for Environmental Monitoring, GIS Applications, and Geology VI, edited by Manfred Ehlers, Ulrich Michel, Proc. of SPIE Vol. 6366 636606-1

Nikolakopoulos, K. G., Kamaratakis, E. K. and Chrysoulakis, N., 2006, SRTM vs ASTER elevation products. Comparison for two regions in Crete, Greece. *International Journal of Remote Sensing*, 27, pp. 4819-4838.

Pan, F., Peters-Lidard, C. D., Sale, M. J. and King, A. W., 2004, A comparison of geographical information systems–based algorithms for computing the TOPMODEL topographic index. *Water Resources Research*, Vol. 40.

Petrie, G. and Kennie, T. J. M., 1990, Terrain Modelling in Surveying and Civil Engineering, 351 pages (London: Whittles Publishing Company)

Rodrıguez-Iturbe, I., 1993, The geomorphological unit hydrograph. In: *Channel network hydrology*, K. Beven and M.J. Kirby (eds), (Chichester: Wiley)

Rodríguez, E., Morris, C. S. and Belz, J. E., 2006, A global assessment of the SRTM performance. *Photogrammetric Engineering & Remote Sensing*, 72, pp. 249-260.

Rojas, R., Velleux, M., Julien, P. Y. And Johnson, B. E, 2008, Grid Scale Effects on Watershed Soil Erosion Models. *Journal of Hydrologic Engineering*, 13, pp. 793–802.

Rosen, P. A., Hensley, S., Joughin, I. R., Li, F. K., Madsen, S. N., Rodriquez, E. and Goldstein, R. M., 2000, Synthetic aperture radar interferometry. Proceedings of the IEEE, 88, pp. 333-382

Comparison of SRTM and ASTER Derived Digital Elevation Models over Two Regions in Ghana – Implications for Hydrological and Environmental Modeling

279

Rosen, P., Eineder, M., Rabus, B., Gurrola, E., Hensley, S., Knöpfle, W., Breit, H., Roth, A., Werner, M., 2001a. SRTM-Mission – Cross Comparison of X and C Band Data Properties. Proceedings of IGARSS, Sydney, Australia, CD.

Shaw, G. and Wheeler, D., 1985, Statistical Techniques in Geographical Analysis (Chichester: Wiley).

Simard, M., Zhang, K., Rivera-Monroy, V. H., Ross, M. S., Ruiz, P. L., Castañeda-Moya, E., Twilley, R. R. and Rodriquez, E., 2006, Mapping height and biomass of mangrove forests in everglades national Park with SRTM Elevation Data. *Photogrammetric Engineering & Remote Sensing*, 72, pp. 299-311

Slater, J. A., Garvey, G., Johnston, C., Haase, J., Heady, B., Kroenung, G. and Little, J., 2006, The SRTM Data "Finishing" Process and Products. *Photogrammetric Engineering & Remote Sensing*, 72, pp. 237-247.

Slater, J. A., Heady, B., Kroenung, G., Curtis, W., Haase, J., Hoegemann, D., Shockley, C. and Kevin, T., 2009, Evaluation of the New ASTER Global Digital Elevation Model. Available online at: http://earth-info.nga.mil/GandG/elevation/ (Accessed 24 June 2011)

Smith, M. P., Zhu, A., Burt, J. E. and Stiles, C., 2006, The effects of DEM resolution and neighbourhood size on digital soil survey. *Geoderma*, 137, pp. 58-69.

Vadon, H., 2003, 3 D Navigation over Merged Panchromatic –Multispectral High Resolution SPOT5 Images. In: *The International Archives of the Photogrammetry, Remote Sensing and Spatial Information Sciences*, Vol. XXXVI, 5/W10

van Zyl, J. J., 2001, The Shuttle Radar Topography Mission (SRTM): a breakthrough in remote sensing of topography. *Acta Astronautica*, 48, pp. 559-565.

Wahba, G., 1990, Spline models for observational data. CNMSNSF Regional conference series in applied mathematics 59. Philadelphia, SIAM.

Walker, J. P. and Willgoose, G. R., 1999, On the effect of digital elevation model accuracy on hydrology and geomorphology. *Water Resources Research*, 35, pp. 2259–2268,

Werner, M., 2001, Status of the SRTM data processing: when will the world-wide 30m DTM data be available? In *Geo Informations systeme*, pp. 6–10 (Heidelberg: Herbert Wichmann/Huthig)

Wilson, J. P., Repetto, P. L. and Snyder, R. D, 2000, Effect of data source, grid resolution and flow-routing method on computed topographic attributes. In: *Terrain Analysis: Principles and Applications*, Wilson, J. P. and Gallant, J. C. (eds). John Wiley & Sons: New York; pp. 133–161.

Wise, S., 2000, Assessing the quality for hydrological applications of digital elevation models derived from contours. *Hydrological Processes*, 14 pp. 1909–1929.

Wolock, D. M. and McCabe, G. J., 1995, Comparison of single and multiple flow direction algorithms for computing topographic parameters, Water Resources Research, 31, pp. 1315–1324.

Yamaguchi, Y., Kahle, A., Tsu, H., Kawakami, T. and Pniel, M, 1998, "Overview of Advanced Spaceborne Thermal Emission and Reflection Radiometer (ASTER)," *IEEE Trans. Geoscience and Remote Sensing*, 36, pp. 1062- 1071.

Zhang, W. and Montgomery, D, 1994, Digital elevation model grid size, landscape representation, and hydrologic simulations. *Water Resources Research*, 30, pp. 1019–1028.

Permissions

The contributors of this book come from diverse backgrounds, making this book a truly international effort. This book will bring forth new frontiers with its revolutionizing research information and detailed analysis of the nascent developments around the world.

We would like to thank Dr. Tommaso Piacentini and Dr. Enrico Miccadei, for lending their expertise to make the book truly unique. They have played a crucial role in the development of this book. Without their invaluable contribution this book wouldn't have been possible. They have made vital efforts to compile up to date information on the varied aspects of this subject to make this book a valuable addition to the collection of many professionals and students.

This book was conceptualized with the vision of imparting up-to-date information and advanced data in this field. To ensure the same, a matchless editorial board was set up. Every individual on the board went through rigorous rounds of assessment to prove their worth. After which they invested a large part of their time researching and compiling the most relevant data for our readers. Conferences and sessions were held from time to time between the editorial board and the contributing authors to present the data in the most comprehensible form. The editorial team has worked tirelessly to provide valuable and valid information to help people across the globe.

Every chapter published in this book has been scrutinized by our experts. Their significance has been extensively debated. The topics covered herein carry significant findings which will fuel the growth of the discipline. They may even be implemented as practical applications or may be referred to as a beginning point for another development. Chapters in this book were first published by InTech; hereby published with permission under the Creative Commons Attribution License or equivalent.

The editorial board has been involved in producing this book since its inception. They have spent rigorous hours researching and exploring the diverse topics which have resulted in the successful publishing of this book. They have passed on their knowledge of decades through this book. To expedite this challenging task, the publisher supported the team at every step. A small team of assistant editors was also appointed to further simplify the editing procedure and attain best results for the readers.

Our editorial team has been hand-picked from every corner of the world. Their multi-ethnicity adds dynamic inputs to the discussions which result in innovative outcomes. These outcomes are then further discussed with the researchers and contributors who give their valuable feedback and opinion regarding the same. The feedback is then collaborated with the researches and they are edited in a comprehensive manner to aid the understanding of the subject.

Apart from the editorial board, the designing team has also invested a significant amount of their time in understanding the subject and creating the most relevant covers. They scrutinized every image to scout for the most suitable representation of the subject and create an appropriate cover for the book.

The publishing team has been involved in this book since its early stages. They were actively engaged in every process, be it collecting the data, connecting with the contributors or procuring relevant information. The team has been an ardent support to the editorial, designing and production team. Their endless efforts to recruit the best for this project, has resulted in the accomplishment of this book. They are a veteran in the field of academics and their pool of knowledge is as vast as their experience in printing. Their expertise and guidance has proved useful at every step. Their uncompromising quality standards have made this book an exceptional effort. Their encouragement from time to time has been an inspiration for everyone.

The publisher and the editorial board hope that this book will prove to be a valuable piece of knowledge for researchers, students, practitioners and scholars across the globe.

List of Contributors

Pavel Raška
Jan Evangelista Purkyně University in Ústí nad Labem, Czech Republic

Enrico Miccadei, Tommaso Piacentini, Francesca Daverio and Rosamaria Di Michele
Dipartimento di Ingegneria e Geologia, Università degli Studi "G. d'Annunzio" di Chieti-Pescara, Chieti scalo (CH), Italia

W. Brian Whalley
Department of Geography, University of Sheffield, Sheffield, United Kingdom

Svetislav S. Krstić
Faculty of Natural Sciences, St.Cyril and Methodius University, Skopje, Macedonia

Cüneyt Baykal and Ayşen Ergin
Middle East Technical University, Department of Civil Engineering, Ocean Engineering Research Center, Turkey

Işıkhan Güler
Yüksel Proje International Co. Inc., Turkey

Gülizar Özyurt and Ayşen Ergin
Middle East Technical University, Civil Engineering Department, Ocean Engineering Research Centre, Turkey

Charles H. Theiling
U.S. Army Corps of Engineers, Rock Island District, Rock Island, Illinois, USA

E. Arthur Bettis
University of Iowa, Department of Geoscience, Iowa City, Iowa, USA

Mickey E. Heitmeyer
Greenbrier Wetland Services, Advance, Missouri, USA

Dávid Lóránt
Károly Róbert College, Hungary

Ronald A. Loughland, Khaled A. Al-Abdulkader, Alexander Wyllie and Bruce O. Burwell
Environmental Protection Department, Saudi Aramco, Saudi Arabia

Joks Janssen
Land Use Planning Group, Wageningen University, Wageningen, The Netherlands

Luuk Knippenberg
Centre for International Development Issues, Radboud University Nijmegen, The Netherlands

Gerald Forkuor
International Water Management Institute, Ghana

Ben Maathuis
Faculty of Geo-Information Science and Earth Observation, University of Twente, The Netherlands

Printed in the USA
CPSIA information can be obtained
at www.ICGtesting.com
JSHW011457221024
72173JS00005B/1113